DANIEL G. PAYNE

Voices in the Wilderness

*American Nature Writing and
Environmental Politics*

University Press of New England
Hanover and London

University Press of New England, Hanover, NH 03755
© 1996 by University Press of New England
All rights reserved
Printed in the United States of America
5 4 3 2 1
CIP data appear at the end of the book

For Anne and our little ones,
Rebecca Caitlin and Emily Rachel

Where is the literature which gives expression to Nature? He would be a poet who could impress the winds and streams into his service, to speak for him; who nailed words to their primitive senses, as farmers drive down stakes in the spring, which the frost has heaved; who derived his words as often as he used them,—transplanted them to his page with earth adhering to their roots; whose words were so true and fresh and natural that they would appear to expand like the buds at the approach of spring, though they lay half-smothered between two musty leaves in a library,—ay, to bloom and bear fruit there, after their kind, annually, for the faithful reader, in sympathy with surrounding Nature.
—Henry David Thoreau, "Walking"

Contents

	Acknowledgments	xi
	Introduction	1
CHAPTER 1	Colonial and Early American Responses to the Wilderness	9
CHAPTER 2	Emerson, Thoreau, and Environmental Reform	29
CHAPTER 3	George Perkins Marsh and the Harmonies of Nature	55
CHAPTER 4	As the Angels Have Departed: John Burroughs and the Religion of Nature	68
CHAPTER 5	The God of the Mountains: The Rhetoric and Religion of John Muir	84
CHAPTER 6	The Days of Wasteful Plenty Are Over: Theodore Roosevelt and His Environmental Legacies	106
CHAPTER 7	Alone in a World of Wounds: The Question of Audience in *A Sand County Almanac*	123
CHAPTER 8	The New Environmentalism and Rachel Carson's *Silent Spring*	136
CHAPTER 9	Monkey Wrenching, Environmental Extremism, and the Problematical Edward Abbey	152
	Conclusion	167
	Index	177

Acknowledgments

My special thanks go to those who guided me through the critical early stages of this project. Marcus Klein (SUNY Buffalo) contributed both his scholarly insight into the literary works I was examining and an unerring sense of the focus and direction of the project as a whole. His mentorship and friendly advice have been—and continue to be—invaluable. Robert Newman (SUNY Buffalo) and John Elder (Middlebury College) were extraordinarily sympathetic readers whose contributions were myriad. I came back to the discussions that Bob and I had on the uses of biography and autobiography many times in the course of my research and writing. I was truly privileged to be able to draw upon John Elder's vast knowledge of American nature writing and have benefited immensely from his kind mentorship as well as his own scholarly work, which I greatly admire. Thanks go also to Dick Ellis (SUNY Buffalo), who provided a historian's perspective, and to John Tallmadge (Union Institute), whose expertise and encouragement meant so much to me as I plodded through seemingly never-ending revisions.

The editorial staff at the University Press of New England have been extraordinarily helpful, especially Mike Lowenthal, who got the project moving toward publication, and David Caffry, whose good humor and steady hand guided me surely through the editorial process.

I would also like to thank my wife, Anne, whose encouragement and support made this work possible, and my parents, Herb and Helen, who always believed. Finally, thanks go to Bob and Liz, Jim, George, JZ, Ken, and Joe, who accompanied me on numerous backpacking trips to the Adirondacks, the White Mountains, and the Green Mountains, all under the guise of "Fieldwork." As John Muir wrote, mere words will not suffice to describe the wonders of nature—one must "come to the mountains and see."

Voices in the Wilderness

Introduction

In a sense, American nature writing has existed ever since European explorers began describing the "New World"—and even long before that if Native American oral traditions are taken into account. The defining aspect of the North American continent was its wilderness, and this was reflected in nearly everything written for at least the first hundred years of the colonial period.[1] Early writers described the land, indigenous plant and animal life, the native inhabitants, the weather, and other natural phenomena in minute detail, leaving abundant documentary evidence of their reactions to the wilderness they confronted. Shaped by European preconceptions (and misconceptions) about nature and the American wilderness, the opinions of these writers about their new surroundings ranged from wasteland to paradise, but one impression stands out as nearly universal—the land needed to be radically transformed in order to make it fit for European settlement. The speed with which this transformation was effected was little short of miraculous. The success of the New England colonists was so complete that just three generations after the landing at Plymouth, Cotton Mather could write: "Never was any Plantation brought into such a considerableness in a space of time so inconsiderable! An *Howling Wilderness* in a few Years [became] a *Pleasant Land*, accommodated with the *Necessaries*, yea, and the *Conveniencies* of Humane Life" (165). Despite this profound change in the landscape, and even despite localized economic disruptions caused by some of these ecological changes, the scope and direction of this metamorphosis wouldn't be seriously called into question for over two centuries.

In *The Comedy of Survival*, Joseph W. Meeker suggests that, throughout history, "Human behavior has generally been guided by presumed metaphysical principles which have neglected to recognize that man is a species of animal whose welfare depends upon successful integration with the plants, animals, and land that make up his environment" (31). There is certainly ample evidence in American ecological history to indicate that this is the case. Instead of a symbiotic integration with the natural world, the predominant American worldview has been, and continues to be, an anthropocentric one that sees the earth as a virtually limitless storehouse from which humankind can extract resources and where it can dispose of waste and alter the landscape with little concern for the ways in which these actions will affect the local or global environment.

Beginning in the nineteenth century, glimmerings of an alternative viewpoint first appeared, developed by writers who saw that nature could serve as an end in itself and was not merely something to be conquered or exploited. Although these early expressions of a nascent ecological sensibility have relatively little in them that we would recognize as distinctively modern, and certainly nothing that might be construed as a political call to arms for environmental reform, they are nonetheless important precursors of the environmental reform movement in America. Writers such as Henry David Thoreau, George Perkins Marsh, and John Burroughs laid the vital groundwork for environmental reform, but not until John Muir did an American nature writer effectively combine esthetic, ecological, economic, and ethical rationales into a persuasive polemic for political change.

Barry Lopez asserts that the genre of nature writing "will not only one day produce a major and lasting body of American literature, but . . . might also provide the foundation for a reorganization of American political thought" (297). I would go even further and suggest that nature writing has already been largely responsible for at least one major change in American political thought, resulting in the land use policies of the Progressive era, particularly during the administration of Theodore Roosevelt. Without the work of writers such as Marsh, Muir, Burroughs, and Roosevelt himself, it is unlikely that sufficient public awareness and support would have existed for the conservation reforms proposed by the Progressives. While it is true that early twentieth-century notions of conservation are no longer thought of as particularly progressive by modern environmentalists, this first high water mark of environmental reform had a lasting influence on the way Americans viewed their land. Issues raised at this time, such as resource management, wilderness preservation, and animal protection, have remained in the forefront of conservation policy debates in the twentieth century, culminating in Aldo Leopold's formulation of a "land ethic" in *A Sand County Almanac*.

Introduction

In the decades following World War II, many nature writers directed their attention toward the dangers posed by nuclear war and radiation, the insidious effects of pathogens such as those contained in pollutants and pesticides, and other environmental hazards unknown to the nature writers of earlier generations. Works such as Rachel Carson's *Silent Spring* increased the public's awareness of the unprecedented ecological damage being caused by certain technologies and inaugurated a second great era of environmental reform. This time, however, many of the issues were far more complex scientifically—and contentious politically—than those the early conservationists had faced, and environmental advocates found themselves confronting a new set of rhetorical challenges and ethical dilemmas. Some of the differences between the early conservation movement and the modern environmental movement have been pointed out by Samuel P. Hays in a recent essay on the development of modern environmentalism. For one thing, he argues, the conservation movement emphasized the efficient use and development of material resources, whereas the environmental movement is concerned with less tangible matters such as quality of life (22). Another, even more profound, difference between the two movements reflects the influence of ecology on the ethical aspects of man's impact on the natural world of which he is himself a part. That is, while conservation has traditionally been human-centered, or anthropocentric, modern environmentalism increasingly assumes a life-centered, or biocentric (and sometimes an even more sweeping geocentric), perspective that takes factors other than human needs and desires into account when making decisions that affect the environment. The ethical implications of these two worldviews are just starting to be examined by critics and, as Barry Lopez intimates, may eventually change the way that humans interact with the natural world. Whether we believe that environmental decision making should be based on the duty we owe to our fellow humans (including those to come) or should proceed, as Paul Taylor has written, "from certain moral relations holding between ourselves and the natural world itself" (12), the ethical underpinnings that inform our relations with the natural world have changed considerably in a relatively short amount of time, and this ethical reassessment has in turn had a marked influence on political discourse in this country.

Perhaps the most remarkable aspect of environmental reform in American politics is the extent to which it has been driven by nature writers. Rarely in American political discourse have writers been so instrumental in molding public opinion—Thomas Paine's *Common Sense* and Harriet Beecher Stowe's *Uncle Tom's Cabin* being two notable exceptions—and the extent to which nature writers have influenced environmental reform in this country may well be unprecedented in American politics. Nature writers have framed the issues of the debate, provided the ethical, ecological, and rhetorical underpinnings of the reformist position, and attracted the mass audience that has formed the basis of both the conservation and environmental movements. While only a few nature writers, notably John Muir and Rachel Carson, were able effectively and simultaneously to combine all three of the key rhetorical tasks—to alert, to inform, and to persuade the reader—each of the writers considered here made an invaluable contribution to the case for environmental reform. A history of environmental reform in America that omitted such names as Muir, Carson, and Aldo Leopold would be a poor history indeed, and I would argue that even those nature writers whose personal agenda did not include political reform were making something of a political statement simply by choosing to write about nature rather than about other, human-centered subjects. While Edward Abbey may be justified in his assertion that it isn't enough merely to write about nature anymore—now one must take direct action to protect it—there is still something to be said, even from a modern political perspective, for a writer who can simply convey a love of nature to others. As Thoreau and Burroughs proved, a nature writer's indirect political contribution, even once or twice removed, can sometimes be far more influential in the long run than the machinations of a law firm full of professional lobbyists.

Moreover, the work of the writers considered in this study continues to be relevant across many disciplines and is fundamentally important to students of environmental literature and politics as well as to environmental advocates. Many—although by no means all—of the issues involved have changed in recent years, but the need for the rhetorical and political sophistication generally displayed by these writers remains a constant. For example, the careful attention to audience that Muir, Leopold, and Carson demonstrate in their work is a rhetorical skill that some of today's environmental advocates, accustomed to preaching only to the converted, appear to have forgotten. This may be attributable in part to the fact that there is now a strong reservoir of public support for environmental protection, whereas writers such as Muir and Leopold once had virtually to create a constituency for conservation through their writings. The recent political gains of those opposed to environmental reform should be a reminder to environmental advocates and writers, however, that political victories can be ephemeral and that the struggle for public opinion and political influence is unremitting. Although the issues and tone of the debate have changed considerably over a century of environmental

Introduction 5

politics, many of the rhetorical lessons provided by the nature writers who formed the vanguard of the reform movement remain relevant. Particularly in an age of sound bites, fuzzy thinking, and demonization, these lessons have an ethical and intellectual dimension that is worthy of our continued consideration.

A Note on Language

Just as American attitudes toward nature have changed since the nineteenth century, so has the language. The concept of "environmentalism" is a comparatively recent one with a broad and complex penumbra of meaning that has evolved mostly since the 1960s. In *The Rights of Nature: A History of Environmental Ethics* Roderick Nash identifies two components to environmentalism: the notion that the abuse of nature is "wrong" and its protection "right"; and a more radical belief that "nature has intrinsic value and consequently possesses at least the right to exist" (9). My use of "environmentalism" here generally reflects this definition, although it is important to note that when I characterize certain passages by early writers as displaying elements of environmental awareness I am anticipating a set of ideas and attitudes that does not coalesce until much later. I have tried not to commit the fallacy of judging writers from another era by contemporary mores; for, although it is a legitimate part of critical inquiry to point out similarities and differences between attitudes then and now, an unwarranted tone of moral condescension frequently creeps into such comparisons. A sense of historical context goes a long way: our society's pollution, profligacy, and dependence on nonrenewable resources will, in all likelihood, be viewed by future generations with the same sense of dismay that we today reserve for many of the practices of the nineteenth century.

Similarly, the terms *conservation* and *preservation* have acquired specialized meanings since the late nineteenth century. The word *conservation* first took on its generally understood meaning—the protection, management, and controlled use of natural resources—around 1875, and the conservation movement originally included both those who emphasized controlled use (such as Gifford Pinchot) and those who emphasized protection (such as John Muir). By 1908 the Pinchot faction of the movement had largely appropriated the term to mean the "comprehensive and well-planned management of natural resources of every character, based on sound ethical and economic grounds" (Huth 185). I have used the word here in that sense, thereby distinguishing it from *preservationism*, the branch of the movement most closely identified with John Muir, which called for a complete (or as close to complete as possible) protection of specific areas for reasons unconnected with their future usefulness as economic resources. Preservationism—with its emphasis on moral, spiritual, esthetic, and biocentric rationales for environmental protection—is more closely related to the mainstream of the modern environmental movement as it has evolved than is conservationism, which is now viewed by many environmentalists as a politically conservative, anthropocentric approach to environmental protection. Other terms have presented similar linguistic challenges; "wise use," for example, a phrase that was originally synonymous with conservation, has acquired additional, more sinister, implications in modern environmental discourse.[2]

Although I generally refer to the writers discussed in this study as "nature writers," I do so reluctantly, fully aware of the deficiencies of the term. A number of contemporary writers commonly referred to as nature writers (including Barry Lopez and Edward Abbey to name just two) have complained, with considerable justification, about being relegated to that particular literary category. Nature writing today encompasses far more than just natural history, and I would suggest that, because we are only beginning to realize that humankind and nature are inextricably linked, both ethically and ecologically, "nature," in all its forms, is *the* theme for our age and one that should not be dismissed simply as the province of "nature writers."

Notes

1. Of course, one society's wilderness is another's home, and the concept of the North American continent as a wilderness would have seemed passing strange to the Native Americans who had lived there for countless generations. As J. Donald Hughes writes in *American Indian Ecology*, North America "was not a wilderness—it was a community in nature of living beings, among whom the Indians formed a part, but not all. There were also animals, trees, plants, and rivers, and the Indians regarded themselves as relatives of these, not as their superiors" (4). See Roderick Nash's *Wilderness and the American Mind* for a concise etymological history of the idea of wilderness (1–7).

See Works Cited sections at the end of each chapter of this volume for full citations of works mentioned.

2. See Kate O'Callaghan, "Whose Agenda for America?" *Audubon* (September–October 1992): 80–91.

Works Cited

Carson, Rachel. *Silent Spring*. Boston: Houghton Mifflin, 1962.
Hays, Samuel P. "Three Decades of Environmental Politics: The Historical Context." In *Government and Environmental Politics: Essays on Historical Developments Since World War Two*, Michael J. Lacey, ed. Washington, D.C.: The Woodrow Wilson Center Press, 1989.
Hughes, J. Donald. *American Indian Ecology*. El Paso: Texas Western Press, 1983.
Huth, Hans. *Nature and the American: Three Centuries of Changing Attitudes*. 1957. Lincoln: University of Nebraska Press, 1990.
Leopold, Aldo. *A Sand County Almanac* (cited as *SCA*). 1949. New York: Oxford University Press, 1966.
Lopez, Barry. "Barry Lopez." In *On Nature*, Daniel Halpern, ed. San Francisco: North Point Press, 1987.
Mather, Cotton. *Magnalia Christi Americana, Books 1 & 2*. Kenneth B. Murdoch, ed. 1702. Cambridge: Belknap Press of Harvard, 1977.
Meeker, Joseph W. *The Comedy of Survival: Studies in Literary Ecology*. New York: Charles Scribner's Sons, 1972.
Nash, Roderick Frazier. *The Rights of Nature: A History of Environmental Ethics*. Madison: The University of Wisconsin Press, 1989.
―――. *Wilderness and the American Mind*. 1967. New Haven: Yale University Press, 1982.
Taylor, Paul. *Respect for Nature: A Theory of Environmental Ethics*. Princeton: Princeton University Press, 1986.

CHAPTER 1

Colonial and Early American Responses to the Wilderness

> We primeval forests felling,
> We the rivers stemming, vexing we and piercing deep the
>   mines within,
> We the surface broad surveying, we the virgin soil upheaving,
> Pioneers! O Pioneers!
> —Walt Whitman, "Pioneers! O Pioneers!" 1865

The most prevalent early reactions to the American wilderness revolved around two powerful and often conflicting responses, hope and fear: hope that the new world would provide a fertile and prosperous new beginning, and fear of a "hideous and desolate wilderness, full of wild beasts and wild men," in William Bradford's well-known phrase. The Puritans added an additional, religious dimension to their "errand into the wilderness," imbuing the ecological changes they worked on the land with a sense of spiritual purpose.[1] Furthermore, as Roderick Nash writes in *Wilderness and the American Mind*, wilderness was not only a formidable threat to the pioneers' very survival, but also had "acquired significance as a dark and sinister symbol. . . . [A]s a consequence, frontiersmen acutely sensed that they battled wild country not only for personal survival but in the name of nation, race, and God. Civilizing the New World meant enlightening darkness, ordering chaos, and changing evil into good" (24). These colonial attitudes, based largely on European preconceptions, would form the basis of colonial and early American policies (or, perhaps more accurately, the absence of such policies) toward nature and the wilderness for over two hundred years. The seemingly limitless forests disappeared at an unprecedented rate, changing ecosystems in unforeseen and sometimes disastrous ways. It was imperative that new responses to nature would be formulated to counteract the initial, outmoded ones. What is perhaps surprising is that the new responses arose from literary and intellectual sources rather than as the result of economic pressures. In an age when the physical survival of colonists often depended on how extensively they could change the land in a short period of time, and when resources were plentiful but labor and capital were not, many of the practices we now characterize as wasteful or foolish become understandable, even justifiable—at least for a time. In addition, as ecological historian William Cronon writes, "Not all the environmental changes which took place after the European settlement were caused by it" (9),[2] and it is probably pointless to argue about whether these changes were historically inevitable. One conclusion *is* inescapable: the American landscape was changed forever in a remarkably short period of time. Lee Clark Mitchell succinctly sums up the central issue here: "Whether one interprets Americans' careless use of the land in social, religious, psychological, or even sexual terms—as say, a compensation for lapsed faith, or a ravishing of the feminine earth—the fact remains that an astonishing portion of the American forest was destroyed" (15).

The first descriptions of the vast American wilderness marvel over the remarkable resources and ecological diversity of the continent and clearly show why profiteers and colonists alike were so optimistic about economic and settlement opportunities in America. Even while discounting some of the more extravagant claims as promotional puffery, we are left with a general impression of unparalleled natural abundance. In *A Description of New England* (1616), John Smith wrote of the "greatness of the timber," the fertility of the land, and the healthful temper of the climate, concluding that New England was the best location for a colony of all the "four parts of the world that I have yet seen" (121). He describes the seemingly infinite plenty of the region and its coastal waters in terms that must have been irresistible to those contemplating opportunities in America:

What pleasure can be more than . . . [for potential settlers] to recreate themselves before their own doors in their own boats upon the sea, where man,

Early Responses to the Wilderness

woman, and child, with a small hook and line, by angling, may take divers sorts of excellent fish at their pleasures? And is it not pretty sport to pull up two pence, six pence, and twelve pence, as fast as you can haul and veer a line? (126).

In *New England's Prospect* (1634), William Wood writes of immense flocks of turkeys, waterfowl, and other game birds, claiming that some hunters had been known to kill forty ducks with one shot (34). He describes the fish, crustaceans, and shellfish of the region as both plentiful and enormous in size, telling of twenty-pound lobsters, oysters a foot in length, and "clamms as bigge as a pennie white loaf" (40). Likewise, the writer(s) of *Mourt's Relation* (presumed to have been William Bradford and Edward Winslow, although the author is identified as George Morton) boasts that "fresh codd in the summer is but course meat with us, our bay is full of Lobsters all the Summer" (135). In *New England's Plantation* (1630), Francis Higginson describes the abundance of edible plants and berries, game, and fish, proclaiming, "Thus wee see both Land and Sea abound with store of blessings for the comfortable sustenance of mans life in New-England" (C). Other writers, such as Thomas Hariot, Robert Johnson, and William Strachey in Virginia, and Thomas Morton in New England, also described the land and its prospects in terms that emphasized its seemingly limitless possibilities. Hariot's *A briefe and true report of the newfound land of Virginia* (1588), for example, is essentially a compendium of the useful resources the colonists had already found (as well as a description of the native inhabitants), and a promise that the unexplored interior of the country "cannot but yeeld many kinds of excellent commodities, which we in our discoverie have not yet seen" (31).

Despite the native abundance to be found in America, the European colonists soon realized that, if they were to live in the manner to which they were accustomed, radical changes in the landscape were necessary. As Roderick Nash writes: "Clearly the American wilderness was not paradise. If men expected to enjoy an idyllic environment in America, they would have to *make* it by conquering wild country" (WAM 26). If colonists had any doubts on this point, the suffering experienced at Jamestown or Plymouth during the first few years after the establishment of those colonies served as a painful lesson. In short, it was inconceivable for the settlers to permit the forest to remain as they had found it, and even some promoters were forced to acknowledge this reality. In *Good News from New England* (1624), for instance, Edward Winslow confirms the reports of New England as a land of plenty, stating, "I will not again speak of the abundance of fowl, store of venison, and variety of fish" (596), but warns prospective settlers that it is not sufficient for "the good creatures of God" to be plentiful, but that means for obtaining them must also be provided for: "Otherwise, as he that walketh London streets, though he be in the midst of plenty . . . is not the better but hath rather his sorrow increased by the sight of that he wanteth, and cannot enjoy" (597).

Throughout the colonies, the desire to transform the wilderness into a garden was a common goal with diverse origins, ranging from the utilitarian to the utopian, and many colonists were eager to transform the landscape into one that resembled the familiar English countryside.[3] For the Puritans, such a transformation was not only necessary from a practical or esthetic sense but was an integral part of their spiritual mission in America. As Peter N. Carroll points out, the Puritans' views of their new world were not uniform, ranging from paradise to wasteland (5), but the wilderness that confronted them was almost universally seen as a challenge to be overcome before the "holy city of God" could be established. Allowing the land to "lie waste" through a lack of cultivation and development would have been an abrogation of their duty to subdue the land as commanded in Genesis 1:28. This duty, they believed, had been ignored by the Indians, and thus they justified their usurpation of Indian lands.[4] Although clearing the forest for planting was a daunting task, writers such as Francis Higginson sought to put a positive spin even on this, reassuring potential settlers that the Indians had already cleared much of the land, and in any event, so many trees made "good living for those that love good Fires." Of course, Higginson had also been the one to rave about the healthful qualities of New England's climate, asserting that "I thinke it is a wise course for all cold complextions to come to take Physicke in New-England for a sup of New-England's Aire is better than a whole draught of old Englands Ale." Whether the air in New England was more healthful than the ale in old England or not, it probably would not have reassured his readers to know that within thirteen months of moving to New England, Higginson had died of fever.

Colonial attitudes about the forest and the Indians were inextricably linked and probably had much to do with the settlers' hostility toward the wilderness. In *New-England's Plantation*, even the eternally optimistic Higginson felt constrained to list some of the downsides of colonial life, which included mosquitos, cold and snowy winters, snakes, and Indians. Although open warfare between the New England colonists and local Indian tribes (some of which had been nearly wiped out by an epidemic just before the landing at Plymouth) didn't flare up in earnest until the Pequot war in 1642, there were periodic scares and skirmishes that reached a peak in 1675 with King Philip's (Metacomet's) war. Tensions between colonists and Indians were a fact of life in most other English colonies, but relations in Virginia had proved to be particularly nettlesome from the start. The Indians explained to Rhode Island Plantation's Roger Williams that they had "*sprung* and *growne* up in that very place, like the very *trees* of the *Wildernesse*" (A4), a version of their genesis that undoubtedly reinforced the settlers' identification of them with the forest and distanced them from the Europeans' biblically based sense of their own divine origins. The colonists' ethnocentric belief that the Indians weren't "using" the land since they had failed to enclose it in the European manner may have been a convenient pretext for usurpation, but it also led inevitably to disagreements between the two sides.[5] When competition for land and game flared into open conflict, forest warfare presented yet another challenge to the colonists, who were faced by a foe for whom the forest provided a significant tactical advantage. As Hector St. John de Crèvecoeur wrote more than a century later in his description of Indian attacks on Mohawk valley settlements during the American revolution, the wilderness was "a harbour where it is impossible to find [the Indians]. It is a door through which they can enter our country whenever they please" (193-94). Deforestation, therefore, not only provided farmland and fuel for "good living for those who like good fires," but was also an early form of ecological warfare that deprived the Indians of both forest cover and the sustenance on which they depended for survival.

The colonists' fear of the wilderness wasn't confined strictly to distrust of the Indians, but extended also to animals such as wolves, "lyons," rattlesnakes, and bears, not to mention a few more fantastic species unknown to modern science. Settlers throughout the colonies systematically destroyed animals they deemed to be noxious, paying particular attention to predators, large and small alike. For instance, John Josselyn describes how the colonists would set out sledges of cod heads in winter to attract foxes. The foxes drawn to this repast would be slaughtered, with the "shooting and killing of *Foxes* [continuing] as long as the moon shineth: I have known half a score kill'd in one night" (267). Wolves were a particular bane to the English livestock and were killed in great numbers. Despite this persecution, the wolves appeared at first to be thriving (probably owing to a rich new food source—the settlers' livestock), and William Wood lamented in 1634 that "there is little hope of their utter destruction, the Countrey being so spacious, and they so numerous" (27). Other large predators likewise seemed impervious to the colonists' not so tender ministrations, and, as late as 1663, diarist John Hull describes the phenomenon of a large number of bears inexplicably pouring out of the wilderness, "so that several hundreds of them were killed by the English in the several parts of this Colony" (210). Other, smaller species such as squirrels and birds thought to damage crops were also hunted ruthlessly, and a bounty was often offered for their destruction, as it was for predators such as wolves and wildcats.[6] Nonetheless, it was probably habitat destruction and not bounty hunting that had the most impact on animal populations, with the possible exception of such commercially desirable species as beaver and other fur-bearers.

With so many compelling reasons for deforestation, the colonists lost little time in clearing away as much of the forest as possible. The speed with which this was accomplished was little short of miraculous and, indeed, was seen by many contemporary observers, particularly in New England, as divinely inspired. In *Wonder-working Providence of Sion's Saviour in America* (1653) Edward Johnson compares New England as it had been before the arrival of the Puritans with what it had been turned into in the short span of thirty years:

this remote, rocky, barren, bushy, wild-woody wilderness, a receptacle for Lions, Wolves, Bears, Foexes, Rockoones, Bags, Bevers, Otters, and all kind of wilde creatures, a place that never afforded the Natives better than the flesh of a few wild creatures and parch't Indian corn incht out with Chestnuts and bitter Acorns, now through the mercy of Christ [has] becom a second England for fertilness. In so short a space, that is indeed the wonder of the world. . . . (210)

While this bleak picture of New England as it had been was somewhat exaggerated in order to provide a moral lesson for those enjoying the fruits of their predecessors' labors, the magnitude of the change wrought on the land was undeniable. Despite the unrelenting war waged on American forests by settlers, however, it is understandable that so few colonial or early American writers expressed misgivings about the possible consequences of wholesale environmental change. For one thing, the changes were seen as beneficial by most observers, who believed that the transformation of the wilderness into a "garden" was desirable. For another, the American forests were so vast that civilization's inroads would seem relatively insignificant for several generations after the initial European colonies were established. So long as settlements were seen as islands in a vast wilderness, the need to preserve the wilderness or protect natural ecosystems would seem unnecessary; only after most of the continent had been settled and wilderness areas were seen as islands in the landscape of a civilized nation did the need to protect them become apparent.

Statements cautioning against profligacy in a land so rich would have seemed incongruous, and the remarkable abundance of the land and its wildlife virtually defined attitudes toward America from the beginning of European colonization. Although clearing land was hard work, as William Cronon points out it was often more profitable to clear and burn new land (in part because of the ready market for potash, a by-product of burning the forest) than it was to remain on old lands (118). The motive for abandoning worn-out farmland for new lands was even more compelling in the south, where the primary cash crop, tobacco, exhausted even virgin farmland in a few years. The perpetual availability of new land encouraged such profligacy as Thomas Jefferson observes in his *Notes on the State of Virginia* (1787): The indifferent state of [agriculture] among us does not proceed from a want of knowledge merely; it is from having such quantities of land to waste as we please. In Europe the object is to make the most of their land, labour being abundant; here it is to make the most of our labour, land being abundant" (125-26). Land hunger and the growing population of the colonies soon prompted what came to be a classically American response to real and perceived land shortages—emigration to the "unclaimed" lands of the frontier. As Cotton Mather fretted, "It was not long before the *Massachusett* Colony was become like an *Hive*, overstocked with *Bees*; and many of the inhabitants entertained thoughts of *swarming* into Plantations extended further into the country" (165). The perpetual availability of new land and resources meant that the day of reckoning for ecologically harmful practices would be put off to an unspecified and unknowable point in the distant future. When game or pelts became scarce from trapping or hunting; when ready supplies of wood disappeared; when soil was exhausted by inefficient farming practices or overgrazing by livestock, there was always "more" just to the west. The size and enormous wealth of the North American continent would delay the consequence of wasteful practices for generations; and without the motivation of immediate scarcity of resources, attitudes were slow to change.

In fact, the exigencies of life in the American wilderness were such that what little in the way of a "conservation ethic" had existed in England was even further diminished once the colonists reached America. English conservation laws served primarily to reserve forests for the crown and game for "gentlemen," although their regulation of such things as acceptable hunting practices, the taking of fish during spawning season, and the protection and reforestation of woodlands reflect a fair degree of ecological sophistication for their time. For many in England, however, such laws were tainted by the stain of class discrimination, thereby associating conservation with elitism—

a confluence that vexes conservationism to a certain extent even now. As the eminent jurist William Blackstone wrote in his *Commentaries on the Laws of England*: "With us in England . . . hunting has ever been esteemed a most princely diversion and exercise" (414). In the colonies, however, such distinctions crumbled under the force of necessity, and for justification the colonists needed to look no further than Blackstone's *Commentaries*, which clearly stated that, in the absence of restrictive laws such as those in England, the "law of nature" applied, whereby "every man, from the prince to the peasant, has an equal right of pursuing, and taking to his own use, all such creatures as are *ferae naturae*, and therefore the property of nobody, but liable to be seized by the first occupant" (411). Blackstone went on to say that this natural right could be "restrained by positive laws enacted for reasons of state, or for the supposed benefit of the community" (411), but colonial legislatures rarely invoked this power. Other restrictions, such as the practice of reserving the best trees for crown use by marking them with a broad arrow, remained on the books but were widely ignored (Cronon 110–11). Albert Cowdrey writes in *This Land, This South: An Environmental History*: "As in the case of wildlife, the freedom to exploit the forests at will was a practical liberty, and its denial a source of irritation between colonists and homeland" (54).

Although it would take many years for the effect of European colonization to become apparent, a few contemporary observers did remark on the changes that were occurring. In 1672, John Josselyn wrote of the flocks of "millions of millions" of passenger pigeons during the fall and spring migrations that extended "for four or five miles, that to my thinking had neither beginning nor ending, length nor breadth, and so thicke that I could see no sun." Although these massive flocks were as yet in no imminent danger of extinction, Josselyn noted that "of late they are much diminished, the English taking of them with Nets" (278). Numerous species of plants and animals disappeared from settled areas, in many instances to be replaced by bothersome European imports. Even the remarkably resilient wolf eventually succumbed to constant pressure and was virtually wiped out in New England by 1800 (Matthiessen 58). In Peter Kalm's *Travels into North America* (1753), the Swedish botanist (a colleague of Linnaeus) comments upon the pervasive profligacy of the colonists, writing: "In a word, the corn-fields, the meadows, the forests, the cattle, &c. are treated with great carelessness by the inhabitants. . . . [T]heir eyes are fixed upon the present gain, and they are blind to futurity" (300). Likewise, the pioneering naturalist William Bartram in his *Travels Through North and South Carolina, Georgia, East and West Florida* (1791) observes

Early Responses to the Wilderness 17

that in the region of Buffalo Lick, Georgia, the animals "are the same which originally inhabited this part of North America, except such as have been affrighted away since the invasion of the Europeans" (30). The buffalo after which the area was named were "not at this day to be seen in this part of the country," he writes, and the elk that had frequented that part of the country were scarce, and only to be found in the Appalachian mountains (30). While some of these ecological changes were so subtle that they went unnoticed, some, such as the disappearance of formerly vast forests, were as starkly evident as the disappearance of the buffalo. In his *Travels in New-England and New-York* (1821), Timothy Dwight explains that the small size of most of the trees in the region was not due to sterility of the soil or other presumed deficiencies of the American climate, as some European scientists, such as Buffon, had postulated, but to the fact that "almost all the original forests of this country [have] been long since cut down" (1:79).

Such dramatic ecological changes did occasionally prompt a reaction from colonial and early American legislatures, usually in response to local wood shortages or other emergencies. John Winthrop's *Journal* indicates that Boston was suffering from a shortage of firewood as early as 1638, and in one entry he describes how several persons experienced frostbite and one died during a wood-gathering expedition to Spectacle Island in the Boston Harbor. Some communities attempted to protect wood supplies by passing laws that restricted private woodcutting on public lands, but such measures were usually ineffectual. There were a few relatively far-sighted settlements in the colonies, such as the Moravians in North Carolina (Cowdrey 35), and William Penn's colony in Pennsylvania, where it was stipulated that in clearing the land for planting, settlers "leave one acre of trees for every five acres cleared, [and] especially to preserve oak and mulberries, for silk and shipping" (*Minutes of the Provincial Council of Pennsylvania*, 28). Even these measures were based solely on economic rationales and, in any event, did little to prevent deforestation. In most cases, when nearby wood supplies disappeared, local communities were left with few options other than a simple market response—that is, importing high-priced wood from forested areas. By 1749 Peter Kalm would report that fuel shortages had hit even Penn's Philadelphia, leading to speculation in wood prices (54), one of the factors that inspired civic-minded Philadelphian Benjamin Franklin to invent a more efficient wood-burning stove.

Despite the depth of the American antipathy toward the wilderness and laissez faire attitude toward resource protection, once the American forest had been transformed into a more pastoral landscape and nature was no longer synonymous with wilderness, a new appreciation for nature became possible. Even some Puritan divines, such as Jonathan Edwards, began to exhibit an interest in nature. While Edwards was a theological student at Yale, he wrote two scientific treatises, "Notes on Natural Science" and "Notes on the Mind," that clearly show the scope of his interest in science, and throughout his life the great preacher of the "New Awakening" seemed to draw particular pleasure and inspiration from his outdoor meditations. In such works as his *Personal Narrative* and *Images or Shadows of Divine Things* (neither of which was published until well after his death) he drew a link between the works of God and nature, a connection that enabled him to appreciate nature's beauty without fear of apostasy. In addition to esthetic pleasure, he found metaphors for his faith in the natural phenomena he observed: "That the works of nature are intended and contrived of God to signify and indigitate spiritual things is particularly evident concerning the rainbow, by God's express revelation" (*Images* 60). A number of other colonial writers, such as William Byrd of Virginia, also show an esthetic and scientific appreciation of nature, at least in their private writings. Like Edwards, Byrd had an interest in natural history (he was a friend of the naturalist/artist Mark Catesby, who occasionally visited Byrd's plantation), and had an eye for natural beauty that is reflected in his journal and in his account of a survey expedition in North Carolina, *The Secret History of the Dividing Line*, which was finally published in 1841.

Eighteenth-century works such as Robert Beverly's *The History and Present State of Virginia* (1705), Mark Catesby's *The Natural History of Carolina, Florida, and the Bahama Islands* (1731), John Bartram's *Observations* (1751), Peter Kalm's *Travels into North America* (1753), and Thomas Jefferson's *Notes on the State of Virginia* (1785), are early indications of the growing interest in the natural history of America that would eventually produce such writers as Thoreau, Burroughs, and Muir. It is William Bartram's *Travels Through North and South Carolina, Georgia, East and West Florida* (1791), however, that most clearly reflects changing sensibilities toward nature.⁷ Although Bartram is not insensible to the dangers and discomforts associated with traveling through the wilderness, he asserts that "scenes of primitive unmodified nature always pleased me" (227), displaying an equanimity toward wilderness that would have been incomprehensible to William Bradford, for one. Even more noteworthy than his esthetic appreciation of wilderness are Bartram's comments on man's ethical duties toward his fellow creatures. While Bartram still believes in the primacy of man as expressed in Genesis, he displays an Assissian belief that this power brings responsibilities, praying:

O Sovereign Lord! Since it has pleased thee to endue man with power, and pre-eminence, here on earth, and establish his dominion over all creatures, may we look up to thee, that our understanding may be so illuminated with wisdom and our hearts warmed and animated, with a due sense of charity, that we may be enabled to do thy will, and perform our duty towards those submitted to our service, and protection, and be merciful to them even as we hope for mercy. (65)

With this sense of duty toward other living creatures guiding him, Bartram's journey becomes one of spiritual growth and awareness as he resolves not to kill even supposedly noxious species such as rattlesnakes (170), and expresses embarrassment at the treatment of Indians by the whites (223). Bartram's *Travels* sold rather poorly in the United States and, as Hans Huth writes in *Nature and the American*, "did not immediately bring about an increase in the general enthusiasm for nature, nor did it induce any prominent writers to emulate its author" (21). However, the impact Bartram had on attitudes toward nature may have been more far-reaching and subtle than his lack of commercial success might indicate. Bartram maintained friendships and correspondences with a number of the leading scientific and literary figures of his time, including Thomas Jefferson, Benjamin Franklin, Peter Kalm, and the ornithologist and poet Alexander Wilson. Additionally, the *Travels* sold quite well in Europe and was extremely popular with a number of the Romantic poets, who would in turn do much to change both European and American views toward nature. Samuel Coleridge was particularly enthused with Bartram's work, referring to the *Travels* as a "series of poems . . . a *delicious* Book" (227).

William Bartram may have appreciated nature "primitive and unmodified by the hand of man," but most of his contemporaries preferred a considerably tamer version. In a passage from his *Travels* that is reminiscent of Cotton Mather, Timothy Dwight hailed the "conversion of a wilderness into a desirable residence for man" and proudly proclaimed that a "forest, changed within a short period into fruitful fields, covered with houses, schools, and churches, and filled with inhabitants, possessing not only the necessaries and comforts, but also the conveniences of life, and devoted to the worship of Jehovah . . . can hardly fail to delight [the mind] of a spectator" (xv–xvi). Dwight's revery of an America covered with "fruitful fields, covered with houses, schools, and churches," was far more typical of his contemporaries than was Bartram's appreciation for "primitive, unmodified nature." As Leo Marx writes, by the end of the eighteenth century, a "fully articulated pastoral ideal of America" had emerged, an ideal that was distinct from the traditional literary setting to which it had previously been relegated (73).

The work of rural agrarianists Thomas Jefferson, Philip Freneau, and Crèvecoeur indicates that the American pioneer/farmer was fast becoming *the* symbolic archetype of the American people. While the wilderness itself was still viewed with antipathy, pioneer life provided esthetic, spiritual, physical and intellectual benefits—not to mention a perfect setting for the growth of American democratic values. Thomas Jefferson is particularly noteworthy in this regard because, perhaps more than any other American writer of the eighteenth century, his writing combines many of the major intellectual sources behind changing American attitudes toward nature. Jefferson's *Notes on the State of Virginia* (1787) is not only indicative of the growing eighteenth-century interest in the natural history of America,[8] but it also exhibits the same growing sense of literary nationalism that would later have such a great impact on changing how Americans saw the world around them. The two forces are sometimes juxtaposed, as in Jefferson's response to the French naturalist Buffon's claim that animals in America were inferior in size, vigor, and quantity to those in Europe. Jefferson's lengthy rebuttal was inspired both by a desire for scientific accuracy and by his defensiveness about how Europeans saw the United States. As Charles A. Miller writes in *Jefferson and Nature*, "insofar as nature symbolized America in its entirety, nature *was* America for Jefferson. His interest in nature and his use of the word are therefore a form of nationalism" (3).[9]

If Americans such as Jefferson were sensitive about real and perceived slights to the national honor, European critics certainly gave them adequate reason to be so. In addition to countering Buffon's theory that species tended to degenerate in America, Jefferson took issue with Abbe Guillaume Thomas Raynal's claim that America had thus far failed to produce any poets or other men of genius. The subject was still a sensitive one fifty years later, when Alexis de Tocqueville wrote:

I readily admit that the Americans have no poets. . . . In Europe people talk a great deal about the wilds of America, but the Americans themselves are insensible to the wonders of inanimate nature, and they may be said not to perceive the mighty forests which surround them till they fall beneath the hatchet. Their eyes are fixed upon another sight: the American people views its own march across these wilds,—drying swamps, turning the course of rivers, peopling solitudes, and subduing nature. (181)

De Tocqueville's assessment, however accurate in the main, failed to note an increased American interest both in literature and in "the wilds of America" that was ongoing even before his travels throughout the country. There were, of course, numerous works, such as Joel

Barlow's elephantine epic poem "The Columbiad," that bear out de Tocqueville's claim,[10] and even the work of poets more sensitive to nature than Barlow often reflect popular nationalist enthusiasms, as a reading of Bryant's "The Prairies" and many of Whitman's poems, including "Song of the Broad-ax," "Song of the Redwood Tree," or "Pioneers! O Pioneers!," amply demonstrates. But ironically, the same nationalistic sentiments that trumpeted the extension of the American nation over the continent were partially responsible for the development of a distinctively American literature that drew upon nature and the wilderness both as sources of inspiration and as the distinguishing feature that set America apart from Europe. Some of this new literary nationalism simply involved transplanting gothic tales to an American setting, as in Washington Irving's "Rip Van Winkle" and "The Legend of Sleepy Hollow," or Charles Brockden Brown's *Edgar Huntley* and *Wieland*, but other writers would take their themes and characters as well from native and natural sources.

William Cullen Bryant was the most prominent of several early American poets, most of whom were heavily influenced by the English Romantics, to make use of native settings and themes. Although Bryant, as was the case with most American writers of his generation, was more concerned with the spiritual and metaphorical side of nature than with natural history per se, he demonstrates a feeling (it is perhaps not specific enough to call it an understanding) for a kind of poetic ecology. In "Thanatopsis" (1817), his first published poem, Bryant explores the spiritual connection between man and nature, emphasizing the comfort that can be drawn, even in "the last bitter hour," from an awareness of the essential unity of all things. To that end he urges the reader to "Go forth, under the open sky, and list / To Nature's teachings, while from all around— / Earth and her waters, and the depths of air— / Comes a small voice" (21). Nature is even more explicitly vested with profound religious implications in some of Bryant's later poems such as "A Forest Hymn" and "To a Waterfowl." Although in his poetry Bryant exhibits more interest in metaphysics than in environmental reform, during his long tenure as editor-in-chief of the New York *Evening Post* (1829–1878), he became a regular editorial supporter of the establishment of national parks and of forest protection.

With the publication of the initial Leatherstocking tale, *The Pioneers* (1821), James Fenimore Cooper became the first American novelist to address the ongoing destruction of America's forests and wildlife. The aging pathfinders Natty Bumppo (Hawkeye) and John Mohegan (Chingachgook) are overtaken by the country's rapid transition from wilderness frontier to civilization as the deer and other game on which they depend have begun to grow scarce. Judge Marmaduke Temple, the founder of Templeton, tries to remedy the situation by imposing stringent game laws, thereby placing an additional burden on the old hunters. It is soon obvious, however, that the scarcity of game is not a result of the hunting practices of Hawkeye and Chingachgook but of the changes that have turned the formerly forested region into farmland and village; or, as Natty puts it, the change is a result of "the money of Marmaduke Temple, and the twisty ways of the law" (291). This painful transition from forest to civilization is ultimately resolved (if it can be called that) by Natty doing the American thing and moving west, but, although Cooper's protagonist can avoid the problem in this manner, Cooper himself realizes that the destruction of the forests created problems that were not so easily resolved.

Cooper structures a number of the most dramatically powerful scenes in *The Pioneers* around his unequivocal condemnation of the profligate waste that often accompanies the transition from forest to civilization. At times he seems eerily perceptive, even prescient, about the extent of the dramatic changes taking place in the American landscape. For instance, in a scene that foretells the extinction of the passenger pigeon a century later, a flock of passenger pigeons passes over the valley and the townspeople turn out to slaughter them in thousands although "None pretended to collect the game, which lay scattered over the fields in such profusion as to cover the very ground with the fluttering victims" (246). In another telling passage, the townspeople seine the lake for fish, catching enormous quantities, although Temple himself realizes that "like all the other treasures of the wilderness, they already begin to disappear, before the wasteful extravagance of man" (260). While Temple professes to abhor waste, and appears to mean well, he participates in the pigeon hunt and is the prime agent in the destruction of the forests around his settlement. During the pigeon hunt, Natty scornfully declines to take part in killing more than he actually needs, causing the chastised Temple to declare that it is indeed time to "put an end to this work of destruction." Still, Natty tries in vain to convince Temple that his actions are primarily responsible for the destruction he condemns: " 'Put an ind, Judge, to your clearings. An't the woods his work as well as the pigeons? Use but don't waste. Wasn't the woods made for the beasts and birds to harbour in?' " (248). However, the very woods are under attack, both from the actions of Temple and from the work of men such as the woodchopper Billy Kirby, who declares "I call no country much improved, that is pretty well covered with trees" (229). While the wilderness ethics of Natty Bumppo and Chingachgook—what Aldo Leopold later called the "go-light" and "one-bullet-one-buck" ethical codes of the pioneer (Leopold, *SCA* 213)—are far more in tune with nature than are those of Temple or the other townspeople, it is obvious that the pathfinders' time has already passed and that a new set of ethics, perhaps those that Temple is struggling to formulate, must take their place if man's "wasteful extravagance" is to be checked. Judge Temple has faith that the law will be the instrument of protecting natural resources, and his rationale is one that prefigures the conservation policies of Theodore Roosevelt: "it is not as ornaments that I value the noble trees of this country; it is for their usefulness. We are stripping the forests, as if a single year would replace what we destroy. But the hour approaches, when the laws will take notice of not only the woods, but the game they contain also" (229). Despite Natty's disdain for "the twisty ways of the law," Cooper makes it clear that forest protection ultimately depends on men such as Temple and the laws they adopt. Cooper was the first American writer to take issue with the old Puritan notion that wilderness was "wasted" land, and to suggest that there might be a point in the not so distant future when Americans might suffer, not from an excess of wilderness, but from an excess of civilization.

The elegiac tone of Cooper's description of the disappearing American wilderness is typical of the first half of the nineteenth century. It is a virtual truism that the first Americans to express an appreciation for the wilderness (such as Cooper) lived in the east, where the wilderness had long since disappeared. Nostalgia for untamed nature didn't extend to westerners, who were still busy taming it, and even while Cooper was struggling to express the effects of the radical changes on the land and its inhabitants in upstate New York, men like Daniel Boone and Davy Crockett were becoming American icons for their work in transforming the wilderness. Astute observers, including a number of explorers and naturalists who traveled through the vast interior of the continent, realized that the transformation now taking place in the western states was progressing at an even more accelerated pace than it had in the east. While searching the Ohio River valley for ornithological specimens, John James Audubon describes in achingly eloquent terms the changes the region had recently undergone:

When I think of these times, and call back to my mind the grandeur and beauty of those almost uninhabited shores; when I picture to myself the dense and lofty summits of the forest, that everywhere spread along the hills, and overhung the margins of the [Ohio River], unmolested by the axe of the settler . . . when I see that no longer any Aborigines are to be found there, and that the vast herds of elks, deer and buffaloes which once pastured on these hills and in these valleys, making for themselves great roads to the several salt-springs, have ceased to exist; when I reflect that all this grand portion of our Union, instead of being in a state of nature, is now more or less covered with villages, farms, and towns, where the din of hammers and machinery is constantly heard; that the woods are fast disappearing under the axe by day, and the fire by night . . . when I see the surplus population of Europe coming to assist in the destruction of the forest, and transplanting civilization into its darkest recesses;—when I remember that these extraordinary changes have all taken place in the short period of twenty years, I pause, wonder, and although I know it all to be fact, can scarcely believe its reality. Whether these changes are for the better or for the worse, I shall not pretend to say." (4–5)

A veritable artistic cottage industry grew up based on the desire of Americans to experience, at least vicariously, the wilderness of the continent before it disappeared entirely. Writers such as Washington Irving (*A Tour of the Prairies*) and Francis Parkman (*The Oregon Trail*) were driven to see the west before it was forever altered, as were painters such as Albert Bierstadt and George Catlin.[11] The romance of the frontier even washed over into politics, as the log cabin became a powerful image that helped numerous candidates attain the presidency, from Andrew Jackson to Abraham Lincoln.[12]

These developments indicate that, more than fifty years before Frederick Turner would propose his frontier thesis of American history, many Americans were at least viscerally convinced that the frontier was what distinguished their nation from Europe. It therefore followed that the complete conquest of the American wilderness might not be an unqualified benefit. As critics such as Roderick Nash and Lee Clark Mitchell have suggested, there was an "undercurrent of apprehension" brought on by the destruction of the American wilderness in the nineteenth century that reveals the ambivalence many Americans felt about their "triumphant conquest of the wilderness" (Mitchell xiv). In relation to Cooper's work, for instance, Nash writes: "The Leatherstocking novels gave Cooper's countrymen reason to feel both proud and ashamed at conquering wilderness" (77). There is little doubt that in just over two hundred years since the first establishment of European colonies on the east coast American attitudes toward nature had changed considerably, if not so considerably as the land itself. If environmental reform was still a concept whose time had not yet come, tentative stirrings in that direction were clearly evident in the first half of the nineteenth century and awaited only the coalescing vision of a new way of looking at the world.

Early Responses to the Wilderness 25

Notes

1. Cecelia Tichi argues that greed and fear were not the sole reasons for the physical changes European settlers worked on the land, that the Puritans, for instance, believed their works were "improving" the land in both practical and esthetic ways. She adds that when we place blame on the colonists for environmental destruction in the modern sense, we confuse effect and intention (2). While I agree with this last point, I believe that Tichi's emphasis on the pejorative "greed" instead of a more neutral phrase such as economic opportunism, sets up a straw man and obscures the very real hope of the Puritans that their colony would not only survive but would turn a profit, as the terms of their charter stipulated. Tichi's point that the Puritans were changing the land to reflect their notions of a divine imperative to do so is important, however, and, while it doesn't extend to most other American colonists, it helps to explain the motivation behind some of the Puritans' actions. See also Cronon, *Changes in the Land* 62 ff., for an excellent discussion of the ways in which the Puritans sought to "improve" the land.

2. The New England Indians' use of fire to clear the forests of underbrush and small trees in order to plant corn, for instance, is well established (Cronon 6). Edward Johnson noted this fact in 1653, writing: "Cutting down of the Woods [the settlers] inclose corne fields, the Lord having mitigated their labours by the Indians frequent fiering of the woods, (that they may not be hindered in hunting Venison, and Beares in the Winter season) which makes them thin of Timber in many places, like our Parkes in England" (85).

3. See Leo Marx's *The Machine in the Garden* for a discussion of this notion of the garden and how it relates to the pastoral ideal in America.

4. As Keith Thomas notes in *Man and the Natural World*, the Puritans weren't the only ones to rationalize land appropriation in this way; the same rationale had been used by Sir John Davies to justify the removal of the Irish from their lands in the early seventeenth century (15).

5. *Mourt's Relation*, for instance, contains a self-serving chapter on "Reasons and Considerations touching the lawfullnesse of removing out of England into the parts of America," which on this basis sought to justify taking land formerly used by the Indians. Roger Williams, on the other hand, opposed this position, claiming that "the *Natives* are very exact and punctuall in the bounds of their Lands . . . notwithstanding a sinfull opinion amongst mauy (sic) that Christians have right to Heathens lands" (95). See also Cronon 55–81.

6. For more on the colonial bounty system, see Lund 32–34; Cronon 132–33, 154; and Cowdrey 48–52.

7. Bartram was the son of famed colonial botanist John Bartram and had accompanied his father on several journeys, providing the illustrations for John Bartram's *A Description of East Florida* (1769).

8. A number of critics have equated the spread of deism among the educated classes in Europe and America with an increased respect for nature among scientists and other intellectuals. See, for example, Arnold Smithline's *Natural Religion in American Literature* and Arthur A. Ekrich, Jr.'s, *Man and Nature in America*.

9. Miller draws attention to the fact that "nature" and "natural" are two of Jefferson's favorite words, appearing with remarkable frequency throughout

his work. As Miller points out, however, there is "a risk to using 'nature' to understand Jefferson," since he attaches (often unintentionally) a number of different, sometimes equivocal meanings to the word (5-6).

10. For a detailed explication of the parallels between Joel Barlow's vision of environmental reform and that of the Puritans, see Tichi 114-50.

11. Particularly since most Americans didn't have the opportunity to see the natural wonders and beautiful vistas of the continent, the work of American landscape painters, especially those associated with the Hudson River school, was of inestimable importance in reordering the ways in which Americans saw and responded to the natural scenery of their native land. A number of commentators, including Hans Huth, Edward Halsey Foster, and Lee Clark Mitchell, have noted the impact that painters such as Thomas Cole and Albert Bierstadt had in glorifying American scenery, and Huth notes that it was the painter George Catlin who made the first public appeal for the creation of national parks in 1833 (135).

12. See John William Ward's *Andrew Jackson: Symbol for an Age* for an insightful discussion of the ways in which Jacksonian era politicians made use of the frontier as a powerful symbol with which to bolster their candidacies.

Works Cited

Audubon, John James. *Delineations of American Scenery and Character*. New York: G. A. Baker & Company, 1926. Essays first published in *Ornithological Biography* (9 vols.) 1831–1839.

Bartram, William. *The Travels of William Bartram*. Francis Harper, ed. New Haven: Yale University Press, 1958.

Blackstone, William. *Commentaries on the Laws of England*. First Worcester edition. Worcester, Mass.: Isaiah Thomas, 1790.

Bryant, William Cullen. *Poetical Works of William Cullen Bryant*. New York: D. Appleton and Company, 1924.

Carroll, Peter N. *Puritanism and the Wilderness: The Intellectual Significance of the New England Frontier 1629–1700*. New York: Columbia University Press, 1969.

Coleridge, S. T. *Collected Works of Samuel Taylor Coleridge, Vol. 12, Marginalia*. George Whalley, ed. Princeton: Princeton University Press, 1980.

Cooper, James Fenimore. *The Pioneers, or the Sources of the Susquehanna: A Descriptive Tale*. 1821. Albany: State University of New York Press, 1980.

Cowdrey, Albert E. *This Land, This South: An Environmental History*. Lexington: University Press of Kentucky, 1983.

Crevecouer, Hector St. John. *Sketches of Eighteenth Century America*. Henri L. Bourdin, et al., eds. New Haven: Yale University Press, 1925.

Cronon, William. *Changes in the Land: Indians, Colonists, and the Ecology of New England*. New York: Hill & Wang, 1983.

De Tocqueville, Alexis. *Democracy in America*. 1835. New York: Mentor Books, 1956.

Dwight, Timothy. *Travels in New-England and New-York* (4 vols.). London: William Baynes & Son, 1823.

Edwards, Jonathan. *Images or Shadows of Divine Things*. Perry Miller, ed. 1948. Westport, Conn.: Greenwood Press, 1977.
Ekrich, Arthur A., Jr. *Man and Nature in America*. New York: Columbia University Press, 1963.
Evans, David. *A History of Nature Conservation in Britain*. New York: Routledge, 1992.
Foster, Edward Halsey. *The Civilized Wilderness: Backgrounds to American Romantic Literature, 1817–1860*. New York: The Free Press, 1975.
Hariot, Thomas. *Thomas Hariot's Virginia*. Theodore Bry, ed. 1588. Ann Arbor: University Microfilms, Inc., 1966.
Higginson, Francis. *New-England's Plantation*. 1630. In *Library of Puritan Writings*, vol. 9, Sacvan Bercovitch, ed. New York: AMS Press, Inc., 1986.
Hull, John. *Diary*. 1663. In *Library of Puritan Writings*, vol. 9, Sacvan Bercovitch, ed. New York: AMS Press, Inc., 1986.
Huth, Hans. *Nature and the American: Three Centuries of Changing Attitudes*. 1957. Lincoln: University of Nebraska Press, 1990.
Jefferson, Thomas. *The Portable Thomas Jefferson*. Merrill D. Peterson, ed. New York: The Viking Press, 1977.
Johnson, Edward. *Johnson's Wonder-Working Providence*. J. Franklin Jameson, ed. 1653. New York: Barnes and Noble, Inc., 1959.
Josselyn, John. *An Account of Two Voyages to New England*. 1675. In *Library of Puritan Writings*, vol. 9, Sacvan Bercovitch, ed. New York: AMS Press, Inc., 1986.
Kalm, Peter. *Travels into North America*. Trans. John Reinhold Forster. 1753–1761. Barre, Mass.: The Imprint Society, 1972.
Leopold, *A Sand County Almanac* (cited as *SCA*). 1949. New York: Oxford University Press, 1966.
Lund, Thomas A. *American Wildlife Law*. Berkeley: University of California Press, 1980.
Marx, Leo. *The Machine in the Garden: Technology and the Pastoral Ideal in America*. New York: Oxford University Press, 1964.
Mather, Cotton. *Magnalia Christi Americana, Books 1 & 2*. Kenneth B. Murdoch, ed. 1702. Cambridge: Belknap Press of Harvard, 1977.
Matthiessen, Peter. *Wildlife in America*. 1959. New York: Viking Books, 1987.
Miller, Charles A. *Jefferson and Nature: An Interpretation*. Baltimore: The Johns Hopkins University Press, 1988.
Minutes of the Provincial Council of Pennsylvania, vol. 1. Philadelphia: Jo. Stevens & Co., printer, 1852.
Mitchell, Lee Clark. *Witnesses to a Vanishing America: The Nineteenth-Century Response*. Princeton: Princeton University Press, 1981.
Morton, George. *Mourt's Relation, or, Journal of the Plantation at Plymouth*. 1622. Ann Arbor: University Microfilms International, 1981.
Nash, Roderick Frazier. *The Rights of Nature: A History of Environmental Ethics*. Madison: The University of Wisconsin Press, 1989.
———. *Wilderness and the American Mind*. 1967. New Haven: Yale University Press, 1982.
Smith, John. *A Description of New England*. 1616. In *Library of Puritan Writings*, vol. 9, Sacvan Bercovitch, ed. New York: AMS Press, 1986.
Smithline, Arnold. *Natural Religion in American Literature*. New Haven: College and University Press, 1966.

Thomas, Keith. *Man and the Natural World.* New York: Pantheon Books, 1983.

Tichi, Cecelia. *New World, New Earth: Environmental Reform in American Literature from the Puritans through Whitman.* New Haven: Yale University Press, 1979.

Ward, John William. *Andrew Jackson: Symbol for an Age.* New York: Oxford University Press, 1955.

Williams, Roger. *A Key into the Language of America.* 1643. Ann Arbor: Gryphon Books, 1971.

Winslow, Edward. *Good News from New England.* 1624. In *Library of Puritan Writings,* vol. 9, Sacvan Bercovitch, ed. New York: AMS Press, 1986.

Winthrop, John. *Winthrop's Journal "History of New England, 1630–1649."* New York: Charles Scribner's Sons, 1908.

Wood, William. *New England's Prospect.* 1634. In *Library of Puritan Writings,* vol. 9, Sacvan Bercovitch, ed. New York: AMS Press, 1986.

CHAPTER 2

Emerson, Thoreau, and Environmental Reform

So much of nature as [the scholar] is ignorant of, so much of his own mind does he not yet possess. And, in fine, the ancient precept, "Know thyself," and the modern precept, "Study nature," become at last one maxim.
—Ralph Waldo Emerson, "The American Scholar," 1837

Shall I not have intelligence with the earth? Am I not partly leaves and vegetable mould myself?
—Henry David Thoreau, Walden, 1855

The impact of Emerson and Thoreau on the way Americans see nature is a subject that has been visited and revisited numerous times, and justifiably so. The work of Thoreau in particular is often used as the touchstone by which other nature writers are judged, and the transcendentalist movement with which Emerson and Thoreau are so closely associated is generally regarded as the first major reordering of the American view of nature. Thoreau's epigrammatical style lends itself well to environmental sloganeering, and his pronouncements on nature and the wild have a talismanic quality to them; as Robert Sayre writes in *Thoreau and the American Indians*: "Today in our time of ecological crisis, people turn to the writings of Henry Thoreau and the speeches and prophecies of great Indian chiefs as if they give a common spiritual support" (ix). Numerous critics, including Roderick Nash, Donald Worster, Max Oelschlaeger,

Hans Huth, and Cecelia Tichi, have given Thoreau (and to a somewhat lesser extent Emerson) substantial credit for changing the way Americans see nature. Despite the immense amount of critical energy that has been directed to the work of these two writers, however, relatively little attention has been given to the specifics of how their views on nature relate to the political side of environmental reform. In other words, were Emerson and Thoreau really as "green" politically as modern environmental sloganeers would have it, or have their sometimes amorphous statements on nature been corrupted to fit into a modern political context?

Keeping in mind that at the time Emerson and Thoreau began their literary careers environmental reform had not yet emerged as a political issue, we can assess their level of environmental awareness in several ways. In the nineteenth century, progressiveness on environmental issues generally ranged from calls for resource management or conservation on the moderate end of the scale to the preservation of wilderness areas in their entirety on the more radical end. Similarly, the environmental ethics expressed in this period range from acknowledgement of only the economic value of natural resources, through other, less tangible, considerations such as aesthetic or spiritual worth, finally arriving at an ecological perspective that views nature as possessing intrinsic worth apart from its value to humans.

In *The Rights of Nature: A History of Environmental Ethics*, Roderick Nash portrays the evolution of environmental ethics as a movement from simple utilitarianism to a belief that "it is right to protect and wrong to abuse nature (or certain of its components) *from the standpoint of human interest*" (9; italics mine). Nash points out that this change invests the argument for conservation with a moral dimension that is absent from the most basic type of utilitarianism, which *doesn't* speak in terms of right and wrong. Yet both positions are essentially anthropocentric, considering nature solely in terms of human needs. A more radical version of environmental ethics, says Nash, holds that "nature has intrinsic value and consequently possesses at least the right to exist. This position is variously called 'biocentrism,' 'ecological egalitarianism,' or 'deep ecology,' and it accords nature ethical status at least equal to that of humans" (9-10). Some commentators, most notably Lynn White, have presented this movement in environmental ethics as a movement away from an axiom of some Christians, derived from Genesis 1:28, that "nature has no reason for existence save to serve man" (1207), which gives additional importance to transcendentalist explorations of the link between God, Man, and Nature. In examining the transcendental movement, I will be looking for three main indicators of environmental awareness: the first is a spiritual affinity with nature that ultimately leads to a rejection of anthropocentrism; the second is the level of ecological awareness (including the effects of human activity) expressed by the writers; the third is the extent to which Emerson and Thoreau took an active role in environmental reform in a political sense.

Most discussions of Emerson's view of nature begin with his groundbreaking essay *Nature* (1836). As for Thomas Jefferson, "nature" has a wide range of meanings for Emerson, and there is little to do with actual, physical nature in *Nature*. As Max Oelschlaeger writes, the conceptual focal point of the essay "is the human soul and God, not nature or the wilderness" (135).[1] Emerson's nature is the "organ through which the universal spirit speaks to the individual, and strives to lead back the individual to it" (*CW* 37), and any distance that we put between ourselves and nature distances us from God as well: "We are as much strangers in nature, as we are aliens from God" (*CW* 39). At the heart of Emerson's conception of God, Man, and Nature is the notion of correspondence, derived from the writings of the Swedish philosopher Emanuel Swedenborg. As Emerson describes it in "The American Scholar," correspondence stands for the idea that "nature is the opposite of the soul, answering to it part for part. One is seal and one is print" (*CW* 55). Through nature then, one could not only learn more about oneself, but could attain a more perfect relation to God. For many of Emerson's contemporaries, *Nature* was not only a release from the ossified dogma of traditional religious teaching but an intellectual declaration of independence as well. In his call for a break from the outdated traditions from the past, and for an "original relation to the universe," Emerson opened the door to a new way of seeing the world. As Walt Whitman writes in a letter to Emerson that was included in the 1856 edition of *Leaves of Grass*: "You have discovered that new moral American continent without which ... the physical continent remained incomplete. ... It is yours to have been the original true Captain who put to sea, intuitive, positive, rendering the first report" (Paul 9).

As Julie Ellison points out, Emerson's use of the idioms of "any and all ideologies, doctrines, disciplines, and professions" makes possible a wide variety of critical interpretations of his work: "His eclectic writing is like the Tar Baby; any interpretation will find surface enough to stick" (77). As a result, despite the absence of any specific language in Emerson's work that expresses a sense of misgiving about the transformation of the American landscape, there is certainly ample evidence of an intuitive ecological awareness in his essays, as even a quick survey of the first few pages of *Nature* reveals:

There is a property in the horizon which no man has but he whose eye can integrate all the parts, that is, the poet.

The lover of nature is he whose inward and outward senses are truly adjusted to each other; who has retained the spirit of infancy even into the era of manhood.

In the woods we return to reason and faith. . . . Standing on the bare ground,— my head bathed by the blithe air, and uplifted into infinite space,—all mean egotism vanishes. I become a transparent eye-ball. I am nothing. I see all. The currents of the Universal Being circulate through me; I am part or particle of God.

The greatest delight which the woods and fields minister, is the suggestion of an occult relation between man and the vegetable. I am not alone and unacknowledged.

Statements such as these have led at least one critic to declare that *Nature* "might even be described as a manifesto for an important strain of Romantic ecological thought" (Worster 103). But while Emerson's writing certainly provides adequate basis for such a reading, there is just as much evidence to make the argument that Emerson's sense of ecology is at best a limited, firmly anthropocentric one. There is a clear understanding in Emerson's work that nature was created by God to benefit man:

All the parts [of Nature] incessantly work into each other's hands for the profit of man. The wind sows the seed; the sun evaporates the sea; the wind blows the vapor onto the field; the ice, on the other side of the planet, condenses rain on this; the rain feeds the plant; the plant feeds the animal; and thus the endless circulations of the divine charity nourish man (*CW* 11).

Emerson restates the old anthropocentric vision of a "kingdom of man over nature" (*CW* 45), holding that nature has no significance of its own outside the intellect of man; or, as he later writes in "The Method of Nature" (1841): "In the divine order, intellect is primary: nature secondary: it is the memory of the mind" (*CW* 123).

Emerson returns to the theme of nature many times, most notably in "The American Scholar" (1837), "The Method of Nature" (1841), "Nature" (Second Series, 1844), and "The Young American" (1844). In these essays, as in *Nature*, there is some evidence to suggest an intuitive grasp of ecological principles; for instance, in "The American Scholar" he speaks of nature as a "web of God . . . [a] circular power returning into itself" (*CW* 54). Just as often, however, he betrays a view of nature much akin to that of his Puritan ancestors.[2] Perhaps the strongest indication of this is in "The Young American," where he delivers a panegyric, heavily laden with nationalistic overtones, on the transformation of "this great savage country" into a land where soon "these wild prairies should be loaded with wheat; the swamps with rice; the hill-tops should pasture innumerable sheep and cattle; [and] the interminable forests should become graceful parks" (*CW* 227). Emerson's often ambivalent thoughts about nature may well be most evident in his journal, where a visit to the Garden of Plants in Paris in 1833 elicits the rapturous resolution, "I will be a naturalist" (*J* 111: 7/13/33); but just a few years later, watching a sunset from the summit of a hill, he writes of his alienation from the "fair impressions" of the scene, lamenting, "I cannot tell why I should feel myself such a stranger in nature" (*JMN* 7:74).

This ambivalence toward the physical world of nature may well be one of the reasons Emerson never makes the leap from his metaphysical musings on nature to a concrete, reformist position regarding environmental politics. As a former minister in an era of declining religious faith, Emerson was looking primarily for a philosophical basis to explain and buttress his own faith, as well as that of others. Therefore, he did not go to nature as a naturalist but as a spiritualist; he was largely uninterested in the hard facts of natural history, and merely went to the woods, as John Burroughs writes, "to fetch the word of the wood-god to men" (221). As has often been pointed out, Emerson's knowledge of natural history was fairly limited, as was his woodmanship, in stark contrast to Thoreau.[3] Although nature and nature study figure prominently in Emerson's essays, his studies of nature were undertaken less for what they would tell him about natural history than for what they might tell him about himself. Indeed, Norman Foerster argues that Thoreau deserves much of the credit for the details of material nature that add substance to Emerson's essays and prevent them from "soaring aloft, in balloon fashion . . . in worlds unrealized" (37).[4] Emerson himself appears to have realized as much; after reading Thoreau's journal in 1863 he wrote in his own, "I find the same thought, the same spirit that is in me, but he takes a step beyond, & illustrates by excellent images that which I should have conveyed in a sleepy generality" (*JMN* 15:352–53). Still, however imperfect Emerson's knowledge of natural history may have been, nature study was central to his philosophy since he believed it was necessary to gain a fuller knowledge of ourselves, serving as a confluence where "the ancient precept, 'know thyself,' and the modern precept, 'study nature,' become at last one maxim" (*CW* 55).

A second reason for Emerson's tacit acceptance of the enormous changes in the American landscape during this period was his implicit belief that these changes were essentially benign. Despite the importance of nature in Emerson's metaphysics, he also has a fascination with technology—"I never was on a coach which went fast enough for me," he writes in his journal (*JMN* 4:296), and in a number of essays he refers to the marvels of nineteenth-century industrialism with evident enthusiasm: "I love the music of the water-wheel; I value the railway; I feel the pride which the sight of a ship inspires; I look on trade and every mechanical craft as education also" (*CW* 120). Several critics, including Sherman Paul and Michael H. Cowan, have stressed the dichotomy between Emerson's views of the country and the city, and although it appears that Emerson's sympathies are tilted in favor of the country, there is always something of the sophisticated urban intellectual in his response to nature. Even in what was probably his most ardent paean to the wilderness, "The Adirondacs" (a poem celebrating a camping trip to those mountains in August 1858), Emerson's gaze is by turns directed inward, to nature's "spiritual lessons pointed home," and east, to the civilized world of Concord and Boston. "We flee away from cities, but we bring / The best of cities with us," writes Emerson, a point reinforced by the emotional highpoint of the poem, the moment when news of the completion of the trans-Atlantic cable is received by the campers:

> Caught from a late-arriving traveler,
> Big with great news, and shouted the report
> For which the world had waited, now firm fact,
> Of the wire-cable laid beneath the sea,
> And landed on our coast, and pulsating
> With ductile fire. Loud, exulting cries
> From boat to boat, and to the echoes round,
> Greet the glad miracle. (*SW* 806)

The rapture with which Emerson hails this "feat of wit, this triumph of mankind," is in stark contrast to Thoreau's dour reaction to the project: "We are eager to tunnel under the Atlantic and bring the old world some weeks nearer to the new; but perchance the first news that will leak into the broad, flapping American ear will be that the Princess Adelaide has the whooping cough" (*Walden* 52).[5] The disparate reactions of Emerson and Thoreau to this event exemplify the difference between their responses to nature and civilization. Emerson's love of nature was a civilized, abstract appreciation, while Thoreau's sprang from the actual woods and fields of Concord. As Donald Worster writes, "Emerson's moral doctrines could not sustain Thoreau for long, for they were aspiring branches that had no roots to support them. They were ideas that were not *soiled* enough" (108). It is difficult to imagine the urbane Emerson claiming that it would be a luxury to "stand up to one's chin in some retired swamp a whole summer day"

(300) as Thoreau did in *A Week on the Concord and Merrimack Rivers*; however, this doesn't diminish the power of his vision of nature—including the power it had for Thoreau himself.

Finally, Emerson's political energies were almost exclusively focused on the issue that he, with most of his countrymen, saw as the central moral dilemma of the age: slavery. As he wrote in his journal in 1851: "We must put out this poison, this conflagration, this raging fever of slavery out of the Constitution" (*JMN* 11:363). Even seemingly unrelated issues, including environmental ones, were affected by the issue of slavery; for instance, expansion into the western territories became even more frenzied after the Compromise of 1820 turned the expeditious settlement of these lands (and their designation as free or slave states) into a political test of wills between north and south. Although the reserved Emerson remained wary of fire-breathing abolitionists such as William Lloyd Garrison, he was convinced beyond a moral certainty that slavery was an intolerable evil, and—particularly after passage of the Fugitive Slave Act in 1850—he gave a number of lectures condemning it. Roderick Nash draws a convincing parallel between nineteenth-century abolitionists and twentieth-century biocentrists, who each "identified an oppressed minority that they think possesses rights and is therefore entitled to liberty from exploitation," and argued for changes in the law from a purely moral stance (*RON* 200). Without making too much of this proposition, it can even be argued that abolition may well have been a sine qua non for the evolution of environmental ethics, since it would probably be difficult to argue that nature had "rights" while a large class of humans were still denied theirs. In any event, it is hard to fault Emerson for not anticipating one reform movement, conservation, when another more pressing one demanded virtually the entire country's attention.

Although other writers would soon go far beyond Emerson in their ideas on environmental reform, the importance of his work in influencing American attitudes toward nature—if not their politics directly—should not be underestimated. As Stewart Udall observes: "If Emerson made no major protest against resource waste during his lifetime, his grand themes nevertheless helped arouse interest in the natural world, and inspired his Concord neighbor, Henry David Thoreau" (48). Emerson's "grand themes" inspired numerous other nature writers and early conservationists as well, including John Muir, who took the aging philosopher on a tour of the Yosemite valley's great sequoia groves in 1871. Emerson's contribution to environmental reform might best be characterized by a metaphor he drew to describe the effect of one of his more abstract essays: "I found when I had finished my new lecture that it was a very good house, only the architect had unfortunately omitted the stairs" (*JMN* 11:327).

Although Thoreau is far more concrete in his use of imagery than Emerson, his fondness for contradiction and paradox makes his work just as problematical—perhaps even more so. As Joyce Carol Oates writes, Thoreau is "the quintessential poet of evasion, paradox, [and] mystery" (*Walden* xv). After reading "A Winter Walk," in 1843, even Thoreau's literary mentor Emerson found occasion to complain of his "old fault of unlimited contradiction" (*J* 313), and numerous modern critics have also pointed out that Thoreau's responses toward nature varied widely and were not always consistent. Like Walt Whitman, whose poetry he admired, Thoreau might dismiss such criticism by proclaiming "Do I contradict myself? / Very well then I contradict myself, / (I am large, I contain multitudes)." Additionally, as James McIntosh points out, from the mid-1840s to the mid-1850s when Thoreau wrote most of his well-known work, including *Walden* and *A Week on the Concord and Merrimack Rivers*, he "was not committed during this period to one view of nature, but was capable of consciously entertaining different and even opposing views, and he felt compelled to present each view in a manner appropriate to it" (211). When we add to this the fact that Thoreau's long neglected later writings, including his journal and his work on seed dispersal and forest regeneration, appear to have been leading toward a synthesis of his views on nature, any ambiguity regarding Thoreau's viewpoint on nature is understandable.

Despite these problems in analysis, however, when Thoreau's published work is read in conjunction with his journal—which some critics are now coming to see as his most remarkable literary achievement—and such late works as "The Succession of Forest Trees" and his recently published manuscripts on seed dispersal, a much clearer view emerges of the direction in which Thoreau was headed.[6] Prior to the recent work on these late writings, many critics held that, after *Walden*, Thoreau's work, with a few exceptions such as his essays on Maine, was in decline. During this period, he devoted increasing attention to his journal, where he compiled voluminous notes on various aspects of natural history. This assessment of Thoreau's work may still hold true from a literary perspective, but when considering the development of his views on nature it has little validity.

Thoreau's reading material and his own writing indicate that he made significant progress toward an ecological understanding of nature in the years prior to his death, and it is this understanding that forms the core of his work on seed dispersal and forest regeneration. In this light we can approach published work such as *A Week on the Concord and Merrimack Rivers, Walden,* and *The Maine Woods* not as the culmination of Thoreau's views on nature but as part of a continuing dialectic in which he was attempting to combine his ideas into a synthesis reconciling philosophy and science. When we approach the works this way, in conjunction with his late work on seed dispersal and forest regeneration as well as his journal, it becomes clear that those who would cite Thoreau as a forerunner of deep ecology are fully justified in doing so.[7]

Thoreau's early views of nature were heavily influenced by Emerson's *Nature,* which had a pivotal effect on Thoreau's intellectual life. Emerson's personal influence was even more important to Thoreau's development; as Walter Harding writes: "It is almost impossible to overestimate the importance of that friendship" (61). The admiring Thoreau wrote in his journal that "Emerson has special talents unequalled. . . . His personal influence upon young persons [is] greater than any man's" (*J* 1:432–33), and it was at Emerson's suggestion that Thoreau began to keep a journal in the first place.[8] It was also Emerson who steered Thoreau in the direction of a literary career, publishing his first efforts in the transcendentalist journal *The Dial.* Perhaps most important, it was essays such as *Nature* and "The American Scholar" that gave an air of intellectual and philosophical legitimacy to Thoreau's nature studies. Thoreau's first essays on natural history, "A Natural History of Massachusetts" (1842), "A Winter Walk" (1843), and "A Walk to Wachusett" (1843), all reflect a heavy Emersonian influence. Even in these essays, however, Thoreau places less emphasis than Emerson on the correlation of nature to the soul, and more on the hard facts of natural history. Although Emersonian notions about correspondence appear in Thoreau's journal as late as 1857 (10:127), it is clear that Thoreau was never merely an imitator of Emerson as some critics, most notably James Russell Lowell, cast him.[9] Joel Porte and James McIntosh have explored the tension in Thoreau's work between Emerson's spiritual vision of nature and his own more experiential version, with McIntosh arguing that the spiritual and the natural "pull against each other in his work" (37). The creative tension between these two forces worked in Thoreau's favor, and eventually led him to fuse his nature studies and his spirituality into a distinctive spiritual ecology in his later work.

By the time Thoreau took up residence at Walden Pond in 1845, his view of nature already differed significantly from that of Emerson. *A*

Week on the Concord and Merrimack Rivers (1849) and *Walden* (1854) each contain some intriguing glimmers of an incipient ecological awareness, derived from Thoreau's field observations and extensive readings in natural history, that go far beyond Emerson's work. Thoreau's firsthand knowledge of natural history; his ambivalence about the technological "progress" of which his countrymen, including Emerson, were so proud; and a religious inclination that merely detoured at transcendentalism before proceeding straight to pantheism—these lie at the heart of many of his comments on man and nature. In a passage that anticipates his own attempt to link science and philosophy, Thoreau criticizes the natural history books of the era as "hasty schedules, or inventories of God's property by some clerk" (97). He asserts that these books lack a spiritual element, a proper understanding of nature as a whole, and that as such they never progress beyond the elementary: "They do not in the least teach the divine view of nature, but the popular view, or rather the popular method of studying nature, and make haste to conduct the persevering student only into that dilemma where the professors always dwell" (97–98).

Thoreau's identification of God with Nature was certainly influenced by Emerson. He goes far beyond Emerson, however, in rejecting Christian dogma, tempering this only somewhat by stating: "It would be a poor story to be prejudiced against the life of Christ because the book had been edited by Christians" (72). He exhibits an undisguised contempt for what he saw as the empty religious convention of New England, which is "full of this superstition, so that when one enters a village the church, not only really but from association, is the ugliest looking building in it, because it is the one in which human nature stoops the lowest and is most disgraced" (77). Thoreau's antagonism toward the church occasionally takes on almost comic overtones, as in one journal entry where he states blandly: "Lectured in basement (vestry) of the orthodox church, and I trust helped to undermine it" (*J* 9:188). Thoreau's hostility toward the church was so well known to his fellow townspeople that Emerson felt obliged to address the issue in his funeral oration, saying:

Whilst he used in his writings a certain petulance of remark in references to churches or churchmen, he was a person of a rare, tender and absolute religion, a person incapable of any profanation, by act or thought. Of course, the same isolation which belonged to his original thinking and living detached him from the social religious forms. This is neither to be censured nor regretted. (*SW* 909)

Not everyone was as tolerant as Emerson when it came to the matter of Thoreau's heterodoxy. A number of reviewers criticized *A Week* on the *Concord and Merrimack Rivers* for what they correctly saw as its author's rejection of conventional Christianity in favor of pantheism. Thoreau had certainly taken no pains to hide his unconventional beliefs, flatly stating that, in his theology, "Pan still reigns in his pristine glory. . . . Perhaps of all the gods of New England and of ancient Greece, I am most constant at his shrine" (65). One of those reviewers, James Russell Lowell, would later clash with Thoreau again over the same subject when, as editor of the *Atlantic Monthly*, he deleted a passage from Thoreau's "Chesuncook" that Lowell found too close to pantheism for his editorial comfort: "It is the living spirit of the tree, not its spirit of turpentine, with which I sympathize, and which heals my cuts. It is as immortal as I am, and perchance will go to as high a heaven, there to tower above me still" (*MW* 122).

Thoreau's unconventional ideas on religion are important primarily because by divorcing himself from the dominant Christian worldview he was able to see the world with the "new eyes" that Emerson had called for in *Nature*. Thoreau's study of nature was in part a substitute for conventional religion, leading him to believe that it was to nature, not to religion, that people must look for answers. "There is more religion in men's science than there is science in their religion," Thoreau writes in *A Week on the Concord and Merrimack Rivers*, and his insistence on concrete detail (for that "fact that may some day blossom into a truth") caused him to reject a teleology based primarily on unquestioning faith. Such faith was intellectually untenable, Thoreau thought, particularly when it was based on the word of "ministers who spoke of God as though they enjoyed a monopoly of the subject" (*Walden* 198). While Thoreau believed that the individual could "see" God, as manifested in nature, he had little patience with those who claimed a special knowledge of the divine: "Tell me of the height of the mountains of the moon, or of the diameter of space, and I may believe you, but of the secret history of the Almighty, and I shall pronounce thee mad" (71).

Thoreau's pantheism would eventually lead him away from anthropocentrism and in *A Week on the Concord and Merrimack Rivers* he questioned assumptions regarding "progress" in a way that demonstrates he was already headed in that direction. While traveling up the Concord River he noted that the salmon, shad, and alewives, formerly abundant in that tributary, had disappeared since the construction of dams, canals, and factories had ended their spawning runs. He tries to take a long view on the matter, musing that:

Perchance, after a few thousand years, if the fishes will be patient, and pass their summers elsewhere meanwhile, nature will have levelled the Billerica dam, and the Lowell factories, and the Grass-ground River run clear again, to be explored by new migratory shoals, even as far as the Hopkinton pond and Westborough swamp. (32)

After considering the subject further, and considering the plight of the "poor shad" struggling upstream to spawn, only to be "met by the Corporation with its dam" (35), Thoreau begins to identify with the fish. In a passage remarkable both for its empathy with nonhuman life and for its anticipation of Edward Abbey's "monkey wrenching" he endorses a more energetic response on behalf of the shad: "I for one am with thee, and who knows what may avail a crowbar against the Billerica dam?" (36).

Realizing that framing the issue in terms of the spawning rights of anadromous fish would probably fail to move most of his fellow New Englanders, Thoreau resorts to a utilitarian argument that foreshadows the rhetorical tactics of later conservationists George Perkins Marsh and John Muir. He points out that valuable farmland lies under the water behind the dam, suggesting that "it would seem that the interests, not of the fishes only, but of the men of Wayland, of Sudbury, of Concord, demand the leveling of that dam" (36). He further suggests that there are other issues of enlightened self-interest involved that man, in the arrogance of his anthropocentrism, ignores at his peril:

> Away with the superficial and selfish phil-*anthropy* of men,—who knows what admirable virtue of fishes may be below low-watermark, bearing up against a hard destiny, not admired by that fellow-creature who alone can appreciate it! Who hears the fishes when they cry? It will not be forgotten by some memory that we were contemporaries. Thou shalt ere long have thy way up the rivers, up all the rivers of the globe, if I am not mistaken. (37)

Playing on the old Puritan distinctions between use and waste, Thoreau returns to the issue of dams in his description of the Merrimack River. The Merrimack, he wryly notes, had been a "mere *waste water*, as it were . . . offering its *privileges* in vain for ages, until the Yankee race came to *improve* them" (86–87). In an age where the benefits of such projects were seldom questioned, Thoreau's disparagement of these "improvements" is, to say the least, unusual. While the question of what his opinion of monkey wrenching would have been is, of course, pure speculation, there are two other items on this score worth considering. First, in 1841 the Concord Lyceum sponsored a debate on the question "Is it ever proper to offer forcible resistance?" and Thoreau and his brother John led the side arguing in the affirmative (Harding 142). Second, in 1859 a number of farmers brought suit against the Billerica mill owners for flooding caused by the dam. Thoreau worked hard to bring supporting material to bear for the farmers' case, and in his journal he quotes with evident approval one of the local farmers, who commented that the river was "dammed at both ends and cursed in the middle" (J 13:149).

Despite the near-mythic status of *Walden* in the genre of nature writing, it probably tells us less about Thoreau's environmental sensibility than any of his other works with the possible exception of *Cape Cod*. In *Walden* Thoreau presents numerous paradoxes and outright contradictions in his views on nature. As Robert D. Richardson, Jr., writes, selective quotations from *Walden* can be misleading, since Thoreau "can be cited on both sides of many issues, and rather easily if one's irony detector is switched off" (237). For instance, in one passage Thoreau writes that trade "curses everything it touches" (70), and in another he writes that commerce is "very natural in its methods. . . . I am refreshed and expanded when the freight train rattles past me" (119). Still, the pantheism that was so evident in *A Week on the Concord and Merrimack Rivers* is just as strong here, and the leveling of his "shrines" in the Walden woods following his departure from the lake brings forth a bitter diatribe on the purity of the lake as opposed to human meanness: "Nature has no human inhabitant who appreciates her. . . . She flourishes most alone, far from the towns where they reside. Talk of heaven! ye disgrace earth" (247). It is also evident that Thoreau's nature studies were starting to result in the ecological insights that are at the heart of his later work. For instance, following his account of how he mapped the pond to determine its depth, he writes:

If we knew all the laws of Nature, we should need only one fact, or the description of one actual phenomenon, to infer all the particular results at that point. Now we know only a few laws, and our result is vitiated, not, of course, by any confusion or irregularity in Nature, but by our ignorance of essential elements in the calculation. (290)

Much of the painstaking effort that Thoreau invested in his attempt to uncover some of these laws of nature is reflected in his journal, which after 1851 is devoted largely to the description and cataloguing of natural phenomena. Thoreau's nature studies led him to believe that "Genesis" was not an antiquated text but an ongoing process; as Henry Beston would write nearly a century later in *The Outermost House: A Year of Life on the Great Beach of Cape Cod*, "Creation is here and now" (173). The earth is not a "fragment of dead history," Thoreau writes, but "living poetry like the leaves of a tree . . . not a fossil earth, but a living earth; compared with whose great central life all animal and vegetable life is merely parasitic" (*Walden* 309).

In *A Week on the Concord and Merrimack Rivers*, Thoreau had proclaimed: "There is in my nature, methinks, a singular yearning toward all wildness" (54); and in *Walden* his rhetoric becomes even more *"extra-vagant"* in its praise of wildness. Cecelia Tichi has argued that "Thoreau's rhetoric of wildness is extravagant but held in check by the smallness or domesticity of his illustrations. . . . His wildness is held in a domestic embrace" (167). There is considerable ground for such a statement. Even in one of the most extravagant passages concerning wildness in *Walden*, where Thoreau talks of seizing and devouring a woodchuck raw for "that wildness which he represented" (210), it must be admitted that the woodchuck is a very small portion of wildness indeed. Such a reading, however, largely overlooks the crucial matter of Thoreau's efforts to create a link between the wild and the civilized. In describing his bean field Thoreau writes: "Mine was, as it were, the connecting link between wild and cultivated fields; as some states are civilized, and others half-civilized, and others savage or barbarous, so my field was, though not in a bad sense, a half-cultivated field" (158).

Beneath Thoreau's facetious description of his half-hearted husbandry is the unifying concept that lies behind all of his declarations in favor of the wild—that wilderness complements civilization and is necessary for the sake of both the individual and society. In "Walking," Thoreau's most extended statement in favor of the wild, he writes, "I feel that with regard to Nature I live a sort of border life, on the confines of a world into which I make occasional and transient forays only" (*EP* 242). As a naturalist, Thoreau undoubtedly knew that border areas are often the most rich and diverse biologically, and it is the border between wilderness and civilization that he finds to be the richest in spiritual and intellectual possibilities. To criticize him for not being wild enough—for living in the woods two miles from Concord rather than living in a remote wilderness for instance—misses the point of his rhetoric of wildness entirely. Thoreau clearly explains the need for such rhetoric in the first paragraph of "Walking": "I wish to make an extreme statement, if so I may make an emphatic one, for there are enough champions of civilization" (*EP* 205). He is convinced that both wilderness and civilization are necessary. As Roderick Nash writes, "for an optimum existence Thoreau believed, one should alternate between wilderness and civilization, or, if necessary, choose for a permanent residence 'partially cultivated country'" (*WAM* 93). In this way, it might well be said that Thoreau anticipates Frederick Turner's frontier thesis by some fifty years.

By the early 1850s Thoreau's belief that the wild was needed as a tonic to civilization had led to a realization that the forests needed to be preserved, especially in New England, where they had been under attack for more than two hundred years. In "Walking" (which was published in 1862, though Thoreau had been presenting versions of it in lectures for a decade), he condemns "man's improvements, so called," such as the cutting down of trees and forests, as actions that "deform the landscape" (*EP* 212). While he doesn't explicitly call for forest preserves, he emphasizes the need to preserve a measure of wildness in both humankind and in the land:

I would not have every man nor every part of a man cultivated, any more than I would have every acre of earth cultivated: part will be tillage, but the greater part will be meadow and forest, not only serving an immediate use, but preparing a mould against a distant future, by the annual decay of the vegetation which it supports. (*EP* 238)

This use of the ecological imagery of soil regeneration to describe the need for a partially "uncultivated" spirit in man also suggests that Thoreau was convinced that humankind was a part of nature, subject to the same general principles as the rest of the natural world. In *A Week on the Concord and Merrimack Rivers* he had already come to the realization that wild nature serves an important societal need: "Even the oldest villages are indebted to the border of wild wood which surrounds them, more than to the gardens of men. . . . Our lives need the relief of such a background, where the pine flourishes and the jay still screams (179). Likewise, in *Walden* he writes: "Our village life would stagnate if it were not for the unexplored forests and meadows which surround it" (317). A few years later, when Thoreau came to realize that the presence of this "border of wild wood" could not be taken for granted, he would take the next step and call for the establishment of forest reserves.

It was during his tenure at Walden Pond that Thoreau took his first trip to the Maine woods, in the summer of 1846. He describes this in "Ktaadn," one of the three essays that comprise *The Maine Woods*.[10] Nowhere is the tension between wilderness and civilization more apparent than in "Ktaadn," often cited as an example of Thoreau's ambivalent feelings toward the wilderness. To a certain extent perhaps this is true, for prior to his travels in Maine Thoreau had no real experience of a forest wilder than the narrow strip of woods around his native village. In *Walden* he had written: "We can never have enough of Nature. We must be refreshed by the sight of inexhaustible vigor, vast and Titanic features. . . . We need to witness our own limits transgressed, and some life pasturing freely where we never wander" (366). In Maine, however, he was taken aback by his first encounter with real wilderness, overwhelmed by the "savage and dreary scenery" of the primeval forest. The old Puritan notion of the "hideous and desolate wilderness" lurks just beneath his prose: "It was but a step on either hand to the grim untrodden wilderness, whose labyrinth of living, fallen, and decaying trees only the deer and moose, the bear and wolf can easily penetrate" (11).

At times the overwhelming fact of wilderness seems to outstrip Thoreau's appetite for wildness: "Talk of mysteries! Think of our life in nature.—daily to be shown matter, to come in contact with it,—rocks, trees, wind on our cheeks! the *solid* earth! the *actual* world! the *common sense! Contact! Contact!* Who are we? *where* are we?" (71). This is one of the most often discussed passages dealing with Thoreau's reaction to wilderness and nature, and Ronald Wesley Hoag argues that it is also "one of the most persistently misinterpreted passages in all of the Thoreau canon" (23). Hoag argues persuasively that critics Sherman Paul and James McIntosh (among many others) have misinterpreted the "Contact!" passage as an instance where Thoreau found "evil" in wilderness. Hoag suggests that the evil Thoreau found in Maine was not the wilderness itself but the actions of humans *in* the wilderness, and that the "Contact!" passage is not indicative of dismay but of awe (24). Even if it is true, as McIntosh has suggested, that this passage is by no means typical, since the stress of the experience temporarily unbalanced Thoreau's relation to nature (179), his response may simply be attributable to the fact that Katahdin and the Maine wilderness were at such a remove from the border area that he preferred. Thoreau might well have been equally overwhelmed had he traveled as deep into "civilization"—the overcrowded tenements of Hell's Kitchen, for example.

By the time of his subsequent trips to Maine in 1853 and 1857, Thoreau was eager for contact with its primeval forests and exhibited none of the trepidation about the "grim, untrodden wilderness" (11) he had displayed in "Ktaadn." By the time he reaches Bangor for his second trip, he writes, "I began to be exhilarated by the sight of the wild fir and spruce tops, and those of other primitive evergreens, peering through the mist in the horizon. It was like the sight and odor of cake to a schoolboy" (*MW* 86). He went to Maine primarily to study the Indians and their relation to the land, and to get a sense of the logging industry's effect on the forest. Appalled by what he saw of the logging industry, he wrote that the mission of the loggers "seems to be, like so many busy demons, to drive the forest all out of the country, from every solitary beaver-swamp and mountain-side as soon as possible" (5). Although the surprising "continuousness" (80) of the forest partially obscured Thoreau's ecological vision in his first trip to Maine, during subsequent trips his attention was drawn to the changes already being worked on the forest, not just by the loggers but, perhaps even more importantly, by the settlers who were changing the very character of the region. He expresses a fear that Maine might soon become as "bare and commonplace" as much of Massachusetts already was, sardonically commenting: "We seem to think that the earth must go through the ordeal of sheep-pasturage before it is habitable by man" (153). He calls attention to how much timber had been stolen from the public lands (145) and how quickly the forests had been shorn of their trees:

The very willow-rows lopped every three years for fuel or powder,—and every sizable pine and oak, or other forest tree, cut down within the memory of man! As if individual speculators were to be allowed to export the clouds out of the sky, or the stars out of the firmament one by one. We shall be reduced to gnaw the very crust of the earth for nutriment (154).

Thoreau's vivid imagery and memorable phrasing gives his angry rhetoric an apocalyptic tone that would be adopted by John Muir some fifty years later in his essays calling for the protection of the Sequoia groves of California. In "The Allegash and East Branch," Thoreau continues his angry denunciation of the "ten thousand vermin gnawing at the base of [the forest's] noblest trees" (228) and answers the loggers' claim that a tree was so big that a yoke of oxen could stand on its stump by saying "the tree might have stood on its own stump, and a great deal more comfortably and firmly than a yoke of oxen can, if you had not cut it down. What right have you to celebrate the virtues of the man you murdered?" (229). The logger "cannot converse with the spirit of the tree he fells, he cannot read the poetry and mythology which retire as he advances" (229), writes Thoreau, exhibiting the same pantheistic concern for the "spirit of the tree" that had caused Lowell to delete a similar passage from "Chesuncook." A passage from Thoreau's journal from 1856 about the felling of an old elm in Concord, pointedly referring to the tree as an "old citizen of the town" and mentioning its "funeral" (J 8:130), further indicates that when Thoreau speaks of the "murder" of a tree his words are not mere hyperbole or anthropomorphization but manifestations of his growing biocentrism.

Similar indications of biocentrism are also evident in Thoreau's comments on moose hunting in "Chesuncook." Thoreau had wanted to watch Joe Aitteon, his Abenaki guide for the 1853 trip, hunt, but after Aitteon succeeded in killing a moose, Thoreau experienced such an acute sense of guilt in spite of his merely peripheral role as an observer that he wrote, "The afternoon's tragedy, and my share in it, as it affected the innocence, destroyed the pleasure of my adventure" (119). The "tragical" experience of the moose hunt—here again he refers to the killing of the moose as "murder" (120)—prompted Thoreau to reflect upon the baseness or coarseness of the motives that led most people into the wilderness. In *Walden* he had written that "[f]ishermen, woodchoppers, and others . . . are often in a more favorable mood for observing her, in the intervals of their pursuits, than philosophers or poets even, who approach her with expectation" (210), but he now suggests that it is the poet, not the lumberman or tanner, who makes the truest use of the woods: "Every creature is better alive than dead, men and moose and pine trees, and he who understands it aright will rather preserve its life than destroy it" (121).[11] The moose hunt, and this last statement regarding the preservation of life, mark something of a turning point in Thoreau's considerations of nature and the wilderness. For the first time, he makes a strong, specific call for preservation, ending "Chesuncook" with a proposal for the establishment of national forest preserves:

Why should not we . . . have our national preserves, where no villages need be destroyed, in which the bear and panther, and some even of the hunter race, may still exist, and not be "civilized off the face of the earth,"—our forests, not to hold the king's game merely, but to hold and preserve the king himself also, the lord of creation,—not for idle sport or food, but for inspiration and our own true recreation? Or shall we, like the villains, grub them all up, poaching on our own national domain? (156)

Thus Thoreau moves from a personal love of nature to a call for public action. Perhaps most significantly, he made this appeal in a public forum, the *Atlantic Monthly*, that probably gave this essay a wider popular currency than anything else he would publish during his lifetime. Most likely, few of the *Atlantic*'s, readers realized that Thoreau's reference to the "lord of creation" in his essay was meant to be ironic, as John Muir's use of the phrase "lord man" would later be.

The most intriguing element of Thoreau's ecological sensibility, and one that is only beginning to receive critical attention, is the way his later nature studies relate to his overall work. Until the last few decades, if his work during this period was considered at all it was usually to disparage it—to cite his "unfortunate" preoccupation with the minutiae of natural history as a classic example of losing the forest for the trees. Walter Harding, for instance, cites a well-known entry from Thoreau's journal, in which he expresses some concern about how "in exchange for views as wide as heaven's cope, I am being narrowed down to the field of the microscope" (*J* 2:406), as evidence that Thoreau felt his work to be changing for the worse (291); and Odell Shepard, editor of *The Heart of Thoreau's Journals*, omitted much of the last several years of Thoreau's journal, claiming that they represented "the gradual conquest of the thinker and poet in Thoreau by the observer" (ix–x). However, such selective quotation can be deceptive. One could just as easily cite passages from his journal that have a completely different connotation: "How sweet is the perception of a new natural fact! suggesting what worlds remain to be unveiled" (*J* 3:441). In *Writing Nature: Henry Thoreau's Journal*, Sharon Cameron argues convincingly that Thoreau was not making random daily notes about nature but was in fact "collecting evidence ('field notes') for an ultimate totality (a 'history of these fields') which at any given moment evades him" (129). Thoreau's attendance to the landscape, as Cameron points out, is an attempt to "read" it (12), an attempt that clearly anticipates a science, ecology, not yet defined as such by scientists.

In *Thoreau's Alternative History*, Joan Burbick suggests that one of the things Thoreau sought to do in his studies of nature was to reconstruct a history of America that privileged nature, an " 'uncivil' history that both challenged the stories of civilization and accepted the implications of geologic time" (1). While there is little evidence to support the notion that Thoreau was "anticivilization," there is no question that Thoreau saw "progress" as a decidedly mixed blessing that had left him with a "tamed and emasculated country" (*J* 8:221). After reading William Wood's *New England's Prospect* (1639) in 1855, Thoreau compared Wood's observations with his own, listing the many plants and animals now either greatly diminished or entirely absent (*J* 7:132–37). The subject was obviously one that rankled, as a year later Thoreau, again using a literary metaphor to describe his ecological studies, writes:

I take infinite pains to know all the phenomena of the spring, for instance, thinking that I have here the entire poem, and then, to my chagrin, I hear that it is but an imperfect copy that I possess and have read, that my ancestors have torn out many of the first leaves and grandest passages, and mutilated it in many places. I should not like to think that some demigod had come before me and picked out some of the best of the stars. I wish to know an entire heaven and an entire earth. All the great trees and beasts, fishes and fowl are gone. The streams, perchance, are somewhat shrunk. (*J* 8:221–22)

It appears, however, that Thoreau sought not only to construct a history of the American forest but in some measure to bring it back, as Donald Worster has suggested in *Nature's Economy: A History of Ecological Ideas* (74). In "The Dispersion of Seeds," a manuscript that includes the lecture on "The Succession of Forest Trees" (delivered before the Middlesex Agricultural Society's meeting on Concord, September 1860), Thoreau predicted that the time would soon come when human action so interfered with the natural regeneration of the forest that "we shall be obliged to plant [trees] as they do in all old countries" (*FIS* 23). Thoreau's studies of how forest succession worked were at the heart of his ideas on reforestation and convinced him that, "when we experiment in planting forests, we [shall] find ourselves at last doing as Nature does. Would it not be well to consult with Nature in the outset?—for she is the most extensive and experienced planter of us all, not excepting the Dukes of Athol" (*FIS* 134).

While Thoreau presented forest regeneration to the Middlesex Agricultural Society in terms of enlightened self-interest—much as he did in the case of the Billerica dam—his main concern was for the forest itself. As James McIntosh points out, behind the straightforward prose of what is essentially a scientific essay lies "the most explicit piece of pantheism Thoreau ever published" (293). The word *Nature* is pointedly and persistently capitalized, and Thoreau refers to the seed, the focal point of his essay, in terms that suggest there is a great deal of religion in his science:

I am interested in the fate or success of every such venture which the autumn sends forth. And for this end these silken streamers have been perfecting themselves all summer, snugly packed in this light chest, a perfect adaptation to this end—a prophecy not only of the fall, but of future springs. Who could believe in prophecies of Daniel or of Miller that the world would end this summer, while one milkweed with faith matured its seeds? (*FIS* 93)

Here then, in his work on seed dispersal, Thoreau finally succeeds in combining science and religion in a spiritual ecology that places mankind and nature in the midst of a continuing act of creation without privileging man's needs over those of nature.

Looking back from here at Thoreau's comments on the desirability of preserving a "border of wild wood" around settled areas, and at his call for national forest preserves in "Chesuncook," these statements seem less like the culmination of his nature studies, and more like steps toward an ecological sensibility that fully meets the criteria indicative of environmental awareness discussed in the beginning of this chapter. The significance of these proposals should not be underestimated, however, as they represent an important, if modest, level of political activity in favor of environmental reform. To this point, there had been only a few similar proposals for the establishment of national parks, the most notable of which was that of the painter George Catlin in 1832. In the 1840s and 1850s others, such as the landscape architects

Andrew Jackson Downing and Frederick Law Olmsted, and newspaper editors William Cullen Bryant and Horace Greeley (who became friends with Thoreau while the latter lived on Staten Island with the family of William Emerson), advocated the creation of city parks, a notion that held much in common with Thoreau's call for a "border of wild wood" near towns. As Thoreau writes in a journal entry for October 15, 1859, each town should have a "park, or rather, a primitive forest, of five hundred or a thousand acres" (13:387), where townspeople could go for recreation and instruction. At the time Thoreau's proposal for national forest preserves appeared in the *Atlantic* in 1858, the idea of such preserves was still a novelty, and, as Roderick Nash points out, prior to John Muir and his Sierra Club most of those who called for the creation of national parks did so for anthropocentric, utilitarian reasons such as water and game supply (*RON* 35–36). This was certainly not the case with Thoreau, who believed that the forests should be preserved for their own sake.

Given Thoreau's adoption of a worldview that clearly placed humankind in nature and not above it, as well as his familiarity with natural history and the rhetorical power of his prose, Thoreau's relative lack of activity in changing public policies toward the environment is somewhat puzzling. While his readership was, at least initially, relatively small—the only two books he published during his lifetime, *A Week on the Concord and Merrimack Rivers* and *Walden*, sold poorly[12] —it seems clear that Thoreau's effectiveness as a spokesperson for nature was curtailed more by his detachment from public affairs than by the limited size of his audience. He displayed a contempt for politics in both his journals and his published work: "Men are degraded when considered as the members of a political organization," he wrote in a journal entry for December 31, 1841 (1:306). And in "Reform and Reformers" he writes: "The great benefactors of their race have been single and singular and not masses of men" (*Reform Papers* 186). Thoreau's strong antislavery sentiments drove him to speak out on behalf of the abolition movement on numerous occasions, but for the most part his journal entries and published essays were devoid of political commentary. This omission is obliquely addressed in a journal entry for November 10, 1851: "A wise man is as unconscious of the movements in the body politic as he is of the process of digestion and the circulation of blood in the natural body." While he admitted that politics was a vital function of society, it "should [be] unconsciously performed, like the vital functions of the natural body" (3:103).

Emerson believed that Thoreau's dissociation from public affairs was unfortunate and said as much in his address at Thoreau's funeral:

Had his genius been only contemplative, he had been fitted to his life, but with his energy and practical ability he seemed born for great enterprise and command; and I so much regret the loss of his rare powers of action, that I cannot help counting it a fault in him that he had no ambition. Wanting this, instead of engineering for all America, he was the captain of a huckleberry-party. Pounding beans is good to the end of pounding empires one of these days; but if, at the end of years, it is still only beans! (*SW* 911)

While it seems odd that Emerson, of all people, would fail to recognize that Thoreau aspired to something spiritually higher than "enterprise," "command," or "pounding empires," Thoreau's societal detachment did limit his direct political contribution to wilderness preservation, and, as John Muir would later show, a spiritual interest in nature and political activism on its behalf were by no means incompatible. While Thoreau might not have conceded that his noninvolvement in public affairs was a fault, he was aware that others counted it one with him, and facetiously claimed in his defense that he had tried it and found "that it does not agree with my constitution. I should not consciously and deliberately forsake my particular calling to do the good which society demands of me, to save the universe from annihilation" (*Walden* 73).

Despite his personal estrangement from public affairs, Thoreau's writings clearly signify a new way of looking at nature as well as a new way of writing about it. In "Walking," Thoreau enunciated a standard for nature writing that probably describes his own work better than anyone else has done:

Where is the literature that gives expression to Nature? He would be a poet that would impress the winds and streams into his service, to speak for him; who nailed words to their primitive sense, as farmers drive down stakes in the spring, which the frost has heaved; who derived his words as often as he used them,—transplanted them to his page with earth adhering to their roots; whose words were so true and fresh and natural that they would appear to expand like the buds at the approach of spring, though they lay half-smothered between two musty leaves in a library,—ay, to bloom and bear fruit there, after their kind, annually, for the faithful reader, in sympathy with surrounding Nature." (*EP* 232)

At its best, Thoreau's nature writing achieves this goal, and if his particular genius did not run to environmental activism, his words would serve to point others in that direction. In a passage from "Walking" that seems to anticipate the effect his works would have on others, he wrote: "Man and his affairs . . . I am pleased to see how little space they occupy in the landscape. Politics is but a narrow field, and that still narrower highway yonder leads to it. I sometimes direct the traveler thither" (*EP* 212–13).

Notes

1. Several other critics have made a similar point: see, for example, Ellison 76, 89; Cowan 183; and Cohen 51.
2. See Cowan 73ff. for an examination of the ways in which Emerson's view of transforming the American wilderness resembles that of the Puritans.
3. See, for example, Sudol 174. Thoreau had some sardonic comments on how Emerson and his companions conducted themselves on their trip to the Adirondacks in 1858, commenting in his journal: "Emerson says that he and Agassiz and Company broke some dozens of ale-bottles, one after another, with their bullets, in the Adirondack country, using them for marks! It sounds rather Cockneyish. . . . Think of Emerson shooting a peetweet (with shot) for Agassiz, and cracking an ale-bottle (after emptying it) with his rifle at six rods! They cut several pounds of lead out of the tree. It is just what Mike Saunders, the merchant's clerk, did when he was there" (J 11:119–20).
4. Hans Huth makes a similar point, acknowledging the importance of Emerson's philosophy in cultivating a new appreciation of nature's worth while stating that "without Thoreau's work in observing nature these ideas would never have received that kind of specific interpretation which made it possible for a widespread public to absorb them" (95).
5. See also K. W. Cameron 45–86. For more on the way in which Emerson's views on technology fit into "The Adirondacs," see Sudol 173–79.
6. See *Faith in a Seed*. See also Sharon Cameron for a close analysis of Thoreau's journal and its place in the overall assessment of his work.
7. Prior to the recent work done on Thoreau's journal and late manuscripts, few critics paid much attention to Thoreau as ecologist. Two of the earliest studies to do so were those of Edward S. Deevey, Jr., and Philip and Kathryn Whitford. In the last decade or so, a number of critics, including Donald Worster (73–74), Roderick Nash (*RON* 36), Max Oelschlaeger (133), David Wagenknecht (130), Richard J. Schneider (119), and Scott Slovic (38), have all touched on this aspect of Thoreau's work to varying extents.
8. Thoreau's first journal entry reads: "Oct. 22 [1837]. 'What are you doing now?' he asked. 'Do you keep a journal?' So I make my first entry to-day" (J 1:3).
9. Thoreau's sometimes antagonistic relationship with Lowell, and Lowell's often petty bashing of Thoreau after the latter's death raise some problematic issues about Lowell and his contribution to progressive nature writing. During his tenure as editor of the *Atlantic Monthly* Lowell was an active sponsor of some of the best nature writing of his day, and wrote a number of creditable pieces himself, including "Humanity to Trees," an 1857 article in the *Atlantic* urging the protection of the California sequoias. Lowell's attacks on Thoreau, safely penned after the latter's death, harm Thoreau's reputation less than they do that of Lowell himself. In 1853, Lowell wrote a book on the Maine woods, *A Moosehead Journal*, of which Edward Halsey Foster aptly writes, "One need only compare Thoreau's *The Maine Woods*, the sympathetic journal of a naturalist, with Lowell's *A Moosehead Journal* (1853), full of learned and irrelevant observations, to understand the difference between Concord and Boston" (181).
10. The essay was first published under the title of "Ktaadn and the Maine Woods" in *Sartain's Union Magazine* in 1848. Thoreau's second trip to Maine

was in 1853, and his essay on that trip, "Chesuncook," was published by the *Atlantic Monthly* in 1858, although it had been delivered in lecture form at the Concord Lyceum late in 1853. His final journey to Maine, described in "The Allegash and East Branch," took place in 1857. This essay wasn't published until after Thoreau's death, when Ticknor and Fields collected his three essays on Maine and published them under the title *The Maine Woods* in 1864.

11. For a detailed explication of Thoreau's attitudes regarding hunting, see Thomas L. Altherr.

12. Of the 1,000 copies of *A Week on the Concord and Merrimack Rivers* printed by Munroe, Thoreau's Boston publisher, only about 225 were sold, and another 75 given away. In 1853 the publisher shipped the remaining 700 volumes to Thoreau, who wryly noted in his journal that he now had "a library of nearly nine hundred volumes, over seven hundred of which I wrote myself" (5:459). *Walden* did better, but did not sell out the initial run of 2,000 copies until 1859.

Works Cited

Altherr, Thomas L. " 'Chaplain to the Hunters: Henry David Thoreau's Ambivalence Towards Hunting." *American Literature* 56:3 (October 1984): 345–61.

Beston, Henry. *The Outermost House: A Year of Life on the Great Beach of Cape Cod*. 1928. New York: Ballantine Books, 1971.

Burbick, Joan. *Thoreau's Alternative History*. Philadelphia: University of Pennsylvania Press, 1987.

Burroughs, John. *Field and Study*. New York: Russell & Russell, 1919.

Cameron, Kenneth Walter. "Emerson, Thoreau, and the Atlantic Cable." *Emerson Society Quarterly*, no. 26 (1962): 45–86.

Cameron, Sharon. *Writing Nature: Henry Thoreau's Journal*. New York: Oxford University Press, 1985.

Cohen, Michael P. *The Pathless Way: John Muir and American Wilderness*. Madison: University of Wisconsin Press, 1984.

Cowan, Michael H. *City of the West: Emerson, America, and Urban Metaphor*. New Haven: Yale University Press, 1967.

Deevey, Edward S. "A Re-Examination of Thoreau's *Walden*." *Quarterly Journal of Biology* (March 1942): 1–11.

Ellison, Julie. *Emerson's Romantic Style*. Princeton: Princeton University Press, 1967.

Emerson, Ralph Waldo. *The Collected Works of Ralph Waldo Emerson: Nature, Addresses and Lectures* (cited as *CW*). Alfred R. Ferguson, et al., eds. Cambridge: Belknap Press of Harvard University Press, 1971.

———. *Emerson in his Journals* (cited as *J*). Joel Porte, ed. Cambridge: Belknap Press of Harvard University Press, 1982.

———. *The Journals and Miscellaneous Notebooks of Ralph Waldo Emerson* (cited as *JMN*) (16 vols.). William H. Gilman, et al., eds. Cambridge: Belknap Press of Harvard University Press, 1960–1982.

———. *The Selected Writings of Ralph Waldo Emerson* (cited as *SW*). Brooks Atkinson, ed. New York: The Modern Library, 1950.

Foerster, Norman. *Nature in American Literature: Studies in the Modern View of Nature*. New York: Russell & Russell, 1923.
Foster, Edward Halsey. *The Civilized Wilderness: Backgrounds to American Romantic Literature, 1817–1860*. New York: The Free Press, 1975.
Garber, Frederick. *Thoreau's Redemptive Imagination*. New York: New York University Press, 1977.
Harding, Walter. *The Days of Henry Thoreau*. New York: Alfred A. Knopf, 1966.
Hoag, Ronald Wesley. "The Mark of the Wilderness: Thoreau's Contact With Ktaadn." *Texas Studies in Language and Literature* 24:1 (Spring 1982): 23–46.
Huth, Hans. *Nature and the American: Three Centuries of Changing Attitudes*. 1957. Lincoln: University of Nebraska Press, 1990.
Lowell, James Russell. *Literary Essays: Among My Books, My Study Windows, Fireside Travels*. Boston: Houghton Mifflin, 1890.
McIntosh, James. *Thoreau as Romantic Naturalist*. Ithaca: Cornell University Press, 1974.
Nash, Roderick. *The Rights of Nature: A History of Environmental Ethics* (cited as *RON*). Madison: University of Wisconsin Press, 1989.
———. *Wilderness and the American Mind* (cited as *WAM*). 1967. New Haven: Yale University Press, 1982.
Oates, Joyce Carol. Introduction. *Walden*. J. Lyndon Shanley, ed. Princeton: Princeton University Press, 1989.
Oelschlaeger, Max. *The Idea of Wilderness*. New Haven: Yale University Press, 1991.
Paul, Sherman. *The Shores of America: Thoreau's Inward Exploration*. Urbana: University of Illinois Press, 1958.
Porte, Joel. *Emerson and Thoreau: Transcendentalists in Conflict*. Middletown, Conn.: Wesleyan University Press, 1965.
Ribbens, Dennis. *The Reading Interests of Thoreau, Hawthorne, and Lanier*. Ph.D. diss., University of Wisconsin, 1969.
Richardson, Robert D., Jr. "The Social Ethics of *Walden*." *Critical Essays on Thoreau's Walden*. Joel Myerson, ed. Boston: G. K. Hall & Co., 1988, 235–48.
Sayre, Robert F. *Thoreau and the American Indians*. Princeton: Princeton University Press, 1977.
Schneider, Richard J. *Henry David Thoreau*. Boston: Twayne Publishers, 1987.
Shepard, Odell, ed. *The Heart of Thoreau's Journals*. Boston: Houghton Mifflin, 1927.
Slovic, Scott. *Seeking Awareness in American Nature Writing*. Salt Lake City: University of Utah Press, 1992.
Sudol, Ronald A. "'The Adirondacs' and Technology." In *Emerson Centenary Essays*, Joel Myerson, ed., 173–79. Carbondale: Southern Illinois University Press, 1982.
Thoreau, Henry David. *Excursions and Poems* (cited as *EP*). 1865. Boston: Houghton and Mifflin, 1906.
———. *Faith in a Seed: The Dispersion of Seeds and Other Late Natural History Writings* (cited as *FIS*). Bradley P. Dean, ed. Washington, D.C.: Island Press, 1993.
———. *The Journal of Henry D. Thoreau* (cited as *J*) (14 vols. bound as 2). Bradford Torrey and Francis H. Allen, eds. New York: Dover Publications, 1962.
———. *The Maine Woods* (cited as *MW*). 1864. Joseph J. Moldenhauer, ed. Princeton: Princeton University Press, 1972.

Reform Papers. Wendell Glick, ed. Princeton: Princeton University Press, 1973.
———. *Walden.* 1855. J. Lyndon Shanley, ed. Princeton: Princeton University Press, 1971.
———. *A Week on the Concord and Merrimack Rivers.* 1849. Carl F. Houde, et al., eds. Princeton: Princeton University Press, 1983.
Tichi, Cecelia. *New World, New Earth: Environmental Reform in America from the Puritans through Whitman.* New Haven: Yale University Press, 1979.
Udall, Stewart L. *The Quiet Crisis and the Next Generation.* 1963. Salt Lake City: Peregrine Smith Books, 1988.
Wagenknecht, Edward. *Henry David Thoreau: What Manner of Man?* Amherst: University of Massachusetts Press, 1981.
White, Lynn, Jr. "The Historical Roots of Our Ecological Crisis." *Science* 155: 3767, 10 March 1967, 1203–1207.
Whitford, Philip and Kathryn. "Thoreau: Pioneer Ecologist and Conservationist." *Scientific Monthly* (November 1951): 291–96.
Worster, Donald. *Nature's Economy: A History of Ecological Ideas.* 1977. New York: Cambridge University Press, 1985.

CHAPTER 3

George Perkins Marsh and the Harmonies of Nature

> Apart from the hostile influence of man, the organic and the inorganic world are . . . bound together by such mutual relations and adaptations as secure, if not the absolute permanence and equilibrium of both . . . at least a very slow and gradual succession of changes in those conditions. But man is everywhere a disturbing agent. Wherever he plants his foot, the harmonies of nature are turned to discords.
> —George Perkins Marsh, *Man and Nature*, 1864

In 1857, the Governor of Vermont, Ryland Fletcher, appointed George Perkins Marsh—a fifty-six-year-old lawyer, businessman, and former congressman—to the position of Fish Commissioner and charged him with investigating the alarming decline of the fish population in the state, with an eye toward replenishing fishing stocks. Despite Marsh's admitted amateur status as a scientist, the report he submitted to the legislature, *Report on the Artificial Propagation of Fish*, was thorough and comprehensive in its analysis of the causes for the decline in the fish population, which Marsh attributed to unwise fishing practices and ecological changes caused by human activity. Although the legislature failed to act upon his recommendations, Marsh's interest in the human effect on the environment continued, leading to the publication in 1864 of *Man and Nature*. This was perhaps the single most important literary contribution to conservation in the nineteenth century, and a work that Lewis Mumford has described as "the fountainhead of the conservation movement" (78).

While Marsh wasn't a nature writer in the general sense of that term, the importance of *Man and Nature* in the development of the conservation movement is such that any discussion of nineteenth-century environmental literature or politics would be incomplete without reference to it. As Paul Brooks writes, "Marsh's highly original study provided both a contrast and a valuable complement to a literature of nature that was based largely on romantic response and esthetic appreciation" (89). Like Thoreau, Marsh was an ecologist even before Ernst Haeckel gave the term its present meaning in 1866, and his was the first major American work to provide scientific rationales for forest preservation. Marsh's arguments in favor of forest and watershed protection were so persuasive and well documented that they would be cited in support of such measures for over fifty years. Marsh had originally intended to call his work "Man the Disturber of Nature's Harmonies," and while that title was eventually discarded, his central thesis remained the same. Man, asserted Marsh, was a destructive force in nature, whose actions resulted in changes that threatened his physical and economic well-being. The short-term pecuniary benefits of clear-cutting forests, for instance, were dwarfed by the long-term damage done to the region's environment. Marsh was no pantheist, and his anthropocentric argument in favor of forest preservation was presented largely in terms of economic considerations, but the publication of *Man and Nature* is a watershed event in the evolution of American attitudes toward nature. The influence of his carefully constructed appeal to "enlightened self-interest" eventually proved to be the decisive element in creating a politically effective conservation movement.

Born in 1801 to a distinguished Vermont family, Marsh's youthful observations and experiences shaped the way he would view the ecological changes that accompanied development. By the late eighteenth century Vermont's landscape had already been radically transformed as an influx of settlers cleared the land for farms and pasturage. In 1791, John Lincklaen described the effect that settlement had already had on the Vermont landscape: "in the Southern parts [of Vermont] where the settlements are older, & the land almost all cleared, the people have cut down almost all the trees, keeping only a small quantity necessary for their own consumption. In the North, where there is more forest, the quantity is more considerable, but no more prized than towards the South (82)." By 1850 most of Vermont was farmland or sheep pasture, and little forested land remained (Lipke 37). Marsh witnessed the rapid destruction of Vermont's forests firsthand, and noted the effect of deforestation on streams and rivers, which were now more prone to flooding during periods of heavy rain and often dried up entirely during droughts. He developed an early scientific interest in nature that was encouraged by his father, a successful lawyer, businessman, and gentleman farmer who had suffered financially by losing a sawmill to a flood on the Queechee River in 1811 (Lowenthal 15),the result of changes brought on by deforestation. Although Marsh maintained a lifelong interest in science, his program of studies while at Dartmouth College concentrated on philosophy and classical languages. Following a short stint as a teacher of the classics at a military academy, he decided to follow his father's example and began to study the law. In 1825 he opened a legal practice in Burlington, Vermont's largest city.

Marsh's cousin James had attended Dartmouth College during the same period George had, and they continued to meet frequently in Burlington after James took a position at the University of Vermont in 1826. Although raised as a Calvinist, George was influenced to some extent by transcendentalism, probably introduced to him by his cousin, whose interpretations of Kant and Coleridge contributed to the development of transcendentalism in America. The Marshes were too traditional in their outlook to approve of what they saw as the more radical aspects of transcendentalism; but the transcendentalist emphasis on nature attracted George to the writings of Thoreau and the poet Jones Very (Lowenthal 17). Hans Huth suggests that Marsh's call for national parks in *Man and Nature* may well have been inspired by Thoreau's proposal for wilderness preserves in "Chesuncook," which appeared in the *Atlantic Monthly* in 1858.

Unlike Thoreau, however, Marsh was active in politics, and in 1843 he was elected (as a Whig) to Congress. Although Marsh's legislative career evinced little of the environmental insight that he would later display in *Man and Nature*, he was instrumental in the creation of the Smithsonian Institution and was named to its fifteen-member Board of Regents in 1847. In the congressional debates over the Smithsonian's charter, dealing with the question of whether the Smithsonian's mission would be limited to the dissemination of applied knowledge or broadened to include pure research, Marsh argued that while the practical benefits of pure research might not be readily apparent, practical applications of science depended upon abstract knowledge as their basis. It was this position that was ultimately adopted by Congress. Prefiguring the rhetorical strategy of *Man and Nature*, Marsh framed the issue in terms of the long-range, tangible interests of the nation, rather than of the abstract value of pure research. Although Marsh lost his congressional seat in 1848, he had campaigned hard and effectively for Zachary Taylor, the newly elected president, who rewarded him with the post of Minister of Turkey. This fortuitous appointment gave Marsh the opportunity to travel extensively in the Levant, during which time he made many of the observations and studies that appeared in *Man and Nature*. To Marsh, the deserts of that once-fertile region were compelling testimony to the long-term effects of humankind on the environment, and they provided him with persuasive evidence that this effect could be catastrophic if left unchecked. The primary "causa causarum" of this environmental degradation, he believed, were the wars and despotism associated with the Roman empire, and "man's ignorant disregard of the laws of nature" (11).

Although the central themes of *Man and Nature* were already beginning to take shape and direct Marsh's inquiries, it would be almost ten years before he organized them in book form. Following his recall from Turkey in 1853, Marsh returned to Vermont, where he suffered a series of financial reverses that compelled him to petition Washington for a new ministerial appointment. One of his failed investments had been with the Vermont Central Railroad, which went bankrupt apparently due to some questionable financial practices. Marsh's bitterness over his substantial loss from the railroad's failure carried over in print to *Man and Nature*, where he railed in a footnote: "Joint stock companies have no souls; their managers, in general, no consciences. . . . In fact, every person conversant with the history of these enterprises knows that in their public statements falsehood is the rule, truth the exception" (51).[1] He was finally offered an appointment as Minister to Italy in 1861, where he continued his environmental studies, particularly of the Alps and Po River region. He began there to work in earnest on *Man and Nature*, finishing the book in 1863.

Marsh's intent in *Man and Nature* was essentially threefold: to identify the effect of man's actions on physical geography; to show the harm inflicted on the natural world by these actions; and to suggest possible corrective measures. Marsh built his argument in *Man and Nature* much as he would have prepared a legal brief, resulting in a work that is ponderous but so precise and well-documented as to be nearly irresistible in its conclusions. Concerning the primary issue, the destructive effect of human-made changes on the earth's physical geography, Marsh argued that widespread harm had already occurred and that immediate remedial actions were called for. He cited specific instances from history where ecological changes had resulted in unforeseen damage, and he suggested that these precedents, when combined with the expert testimony of European authorities on geography and forestry, led to only one possible conclusion: humankind had been guilty of disturbing the essential harmonies of nature, and, unless corrective measures were taken, the result would be widespread ecological catastrophe.

Marsh began his case by referring to the Roman empire, arguing that the geographical deterioration of that region, owing to human actions, was as much a cause of the decline of the empire as was "the direct violence of hostile human force" (10). While he never refers directly to Edward Gibbons's work regarding the decline of the Roman empire, Marsh's geocentric theories explaining the collapse of that great civilization countered Gibbon's anthropocentric explanation that religious and sociopolitical forces were the cause of Rome's decline. Marsh suggested that while people could work tremendous changes on the land, at some point they would lose control over the processes they had set in motion, and the resulting ecological changes would in turn produce economic and political upheaval. In succeeding chapters he examined the "transfer, modification, and extirpation" of plant and animal species and devoted a series of chapters to humankind's effect on the woods, waters, and sands. The longest chapter, "The Woods," comprising over a third of the book, is largely concerned with the damage caused when woodlands are indiscriminately destroyed; this is the section that provided much of the material for a later generation of foresters and conservationists seeking to preserve America's forests.

While Marsh was not the first American writer to publicly call for the establishment of protected woodlands (the artist George Catlin had done so in 1832), he was the first to provide a comprehensive, scientific basis for doing so. Marsh's work detailed the ways in which woodlands influenced the local ecology, even in communities a considerable distance from the forest. Drawing on his own experience in Vermont, his extensive travels in Europe and the Levant, and the work of European foresters, he traced the critical role of forests in protecting watersheds and preventing flooding and erosion. He reviewed the available evidence and concluded that forests moderated extremes in temperature and preserved soil moisture, and he suggested that the destruction of woodlands resulted in a diminution of rain and dew. He also raised the possibility that forests might actually act to produce rain, but cautioned that the evidence here was inconclusive (170).

The evidence *was* conclusive, however, as to what the consequences were when forests were destroyed, and Marsh presented those consequences in stark terms: "With the disappearance of the forest, all is changed . . . the climate becomes excessive, and the soil is alternately parched by the fervors of summer, and seared by the rigors of winter" (187). Flooding becomes commonplace, he asserted, and once-fertile lands erode and become barren while "the rich organic mould which covered them, now swept down into the dank low grounds, promotes a luxuriance of aquatic vegetation that breeds fever, and more insidious forms of mortal disease, by its decay, and thus the earth is rendered no longer fit for the habitation of man" (187). Marsh referred to specific examples from history to illustrate this near-apocalyptic vision of environmental destruction, citing the desertification of large portions of the Levant, the floods and landslides of the Alps, and the pestilence of the Po River valley as instances where deforestation and overgrazing by domestic livestock had made once prosperous regions nearly uninhabitable.

Central to Marsh's consideration of what he termed "physical geography" was the notion of interconnectedness, a natural harmony dependent upon a balance between both organic and inorganic components in nature: "Thus all nature is linked together by invisible bonds, and every organic creature, however low, however feeble, however dependent, is necessary to the well-being of some other among the myriad forms of life with which the Creator has peopled the earth" (96). While the usefulness of some species was apparent—or became evident after their destruction[2]—humankind was destroying (sometimes inadvertently) countless organisms, such as coral builders, whose usefulness was not yet understood (110).

Although there were others who were becoming aware of a link between living organisms and their environment (Thoreau, for one), as David Lowenthal notes, at the time of Marsh' work "it was novel to consider these ecological relationships as part of science at all" (258). Most significantly, Marsh's work was the first to emphasize man's actions in the equation and to amass hard evidence on the effects of man-made changes in the web of ecological relationships. Roderick Nash writes that Marsh "stood practically alone among his contemporaries in bringing a rudimentary scientific analysis to man-land relations" (194). In this he is a direct predecessor of such literary ecologists as Aldo Leopold and Rachel Carson.

Despite his condemnation of humankind's wasteful and destructive practices, Marsh was still firmly anthropocentric in his orientation. Underlying his study was the premise that "man is, in both kind and degree, a power of a higher order than any of the other forms of animated life, which like him, are nourished at the table of bounteous nature" (3). Ironically, Marsh argued that it was man's capacity for destructiveness that served as proof of this proposition. Man alone among creatures could upset nature's harmonies, thereby demonstrating that, "though living in physical nature, he is not of her, that he is of more exalted parentage, and belongs to a higher order of existence than those born of her womb and submissive to her dictates" (37). In his biography of John Muir, Michael Cohen points out that this central paradox in Marsh's argument was only slightly more enlightened than the traditional version of anthropocentrism as initially defined in Genesis 1:28 (21), and to a large extent this is true. Marsh believed that it was appropriate for humans to mold nature in order to serve legitimate needs; unfortunately, he writes, mankind has gone beyond his justifiable struggle to meet legitimate needs by shaping nature, and has carried on an "almost indiscriminate warfare upon all forms of animal and vegetable existence around him" (40). In short, while Marsh was certainly no deep ecologist, his stern admonition that "man has too long forgotten that the earth was given to him for usufruct alone, not for consumption, still less for profligate waste" (36), was still far ahead of its time, and it is in that context that his work should be considered.

Just as Marsh's understanding of ecological relationships was prescient, so was his plan for the role conservation could play in preserving nature's harmonies. After presenting the question of humankind's destructiveness in apocalyptic terms, he posed the final and most vital of his three inquiries, the question of what humankind could do to alleviate the environmental damage it had caused:

It is, as I have before remarked, a question of vast importance, how far it is practicable to restore the garden we have wasted, and it is a problem on which experience throws little light, because few deliberate attempts have yet been made at the work of physical regeneration, on a scale large enough to warrant general conclusions in any one class of cases. (353)

However, there were a few instances where restorative measures had been attempted. In Tuscany, for instance, an area Marsh had studied closely during his diplomatic tenure in Italy, he believed it was still possible to take restorative measures to control the damage humans had already worked on the land. Marsh asserted that preservation of existing woodlands and reforestation of severely afflicted areas would mitigate man's tendency "to insure the final exhaustion, ruin, and desolation of every province of nature which he has reduced to his dominion" (352).

Marsh believed that, although humankind's destructiveness had resulted in widespread ecological damage, it still lay within our power to reverse the process. However, as with his successful defense of the Smithsonian's mission to advance pure science, Marsh also realized that any argument to protect natural resources would have to be framed in utilitarian and practical, rather than moral or aesthetic terms. Although the extent and effect of deforestation was not yet as evident in North America as it was in Europe, the American forest was receding at an unprecedented rate. Marsh drew attention to the precipitous decline in America's forests, warning that "the vast forests of the United States and Canada cannot long resist the improvident habits of the backwoodman and the increased demand for lumber" (257). If the trend was not reversed, he feared, Americans would soon feel the consequences of their "improvident habits" as ever more severe ecological catastrophes would follow deforestation. Marsh's proposed remedy was based in part on an examination of European legislation on sylviculture, which he found generally divided "into two great branches—the preservation of existing forests, and the creation of new" (259). The second half of Marsh's proposed remedy for the problem of deforestation—the creation of new woodlands—was even more problematical than his section on forest preservation. The cultivation of timber was a new concept in America, but Marsh discussed it at length, again drawing upon information gathered from European foresters. He outlines the advantages to be gained from cultivating and managing new forests, which not only would mitigate the consequences of previous deforestation but would also provide the economic benefit of producing sustainable yields of timber that would not harm the forest. He condemns clearcutting as a "wasteful economy" that destroys the "mixed character of the forest—in many respects an important advantage, if not an indispensable condition of growth" (270).

Throughout *Man and Nature* Marsh emphasizes the practical benefits of his program. While a government—local, state, or federal—"endowed with an intelligent public spirit" might act to preserve and replant woodlands for intangible or long-term benefits, Marsh did not want to depend upon such a spirit. The problem of short-sighted forest management was even more acute for privately held woodlands. The dangers of deforestation might not become readily apparent for years, and even if a landowner understood the damage he had caused to the land, in a society where land was cheap and plentiful the likely solution would be to move on rather than invest in a costly reforestation project that was unlikely to pay dividends in the planter's lifetime: "The growth of arboreal vegetation is so slow that, though he who buries an acorn may hope to see it shoot up to a miniature resemblance of the majestic tree which shall shade his remote descendants, yet the longest life hardly embraces the seedtime and the harvest of a forest" (278).

While government policies could encourage forest preservation, Marsh believed that "the planter of a wood must be actuated by higher motives than those of an investment the profits of which consist in direct pecuniary gain to himself or even to his posterity" (278). In calling for a new relationship between mankind and the land based on "higher motives," Marsh expressed a need to educate the American people concerning the ethical reasons for forest protection: "the American people must look to the diffusion of general intelligence on this subject, and to the enlightened self interest, for which they are remarkable, not to the action of their local or general legislatures" (259). In *A Sand County Almanac*, Aldo Leopold would outline an ethical approach to land use very similar to Marsh's, writing that "a system of conservation based solely on economic self-interest is hopelessly lopsided. . . . An ethical obligation on the part of the private owner is the only visible remedy for these situations" (251).

With the possible exception of Oregon, Marsh doubted that any American state had more woodland at the moment that it ought permanently to protect. Most states had long since sold off to private interests woodlands that might otherwise have come under governmental protection—and all too often this had resulted in their destruction. He conceded that, even if the states had placed these lands under their direct protection, forested lands were so little valued in the United States that neither public nor private preserves were likely to be respected. Thoreau had made a similar observation in "Chesuncook".[3]

Unfortunately, for all the care Marsh took in documenting his research and formulating his arguments, the American people would require other, more accessible writers for "the diffusion of general intelligence" on the need for forest preservation. The complex nature of Marsh's arguments, his exhaustive documentation, and his idiosyncratically didactic style make for a work that often borders on the tedious. Cognizant of the sheer weight of the scientific, historical, and geographical material upon which his arguments were based, Marsh was frequently obliged to stop and summarize his analysis. Cecelia Tichi writes that *Man and Nature* is "a lament in statistics" (219). Likewise, Paul Brooks writes that "no one could ever say of *Man and Nature*, as Howells did of Burroughs' *Wake-Robin*, that it is a sort of summer vacation to turn its pages. Marsh is not concerned about straining the reader's intellect" (90). Despite Marsh's claim that he was writing for a general, albeit intelligent audience, his turgid, legalistic style made it unlikely that he would reach any but a select few.

For all its literary shortcomings, *Man and Nature* did manage to find a receptive and influential audience. William Cullen Bryant quoted the book at length in an 1864 editorial in the New York *Evening Post* on the uses of trees (Huth 169), and it was often cited by proponents of the nation's first two large wilderness parks, Yellowstone (1872), and New York State's Adirondack Park (1885). Selling approximately a thousand copies at its initial publication, over the next thirty years *Man and Nature* was reprinted several times. Marsh's work had its greatest impact among succeeding generations of foresters and preservationists who took his strictures to heart and brought his persuasive arguments to the public debate over conservation. Among those influenced by Marsh was John Muir, who owned an annotated copy of *Man and Nature* and cited it in support of his own efforts to save California's big trees. Although some preservationists—including Muir— disagreed with the anthropocentrism at the heart of *Man and Nature*, its findings were still immensely useful to their efforts because, as Roderick Nash suggests, the book "made protecting wilderness compatible with progress and economic welfare" (105).

Marsh claimed to be writing for a general audience, believing that the "common observation of unschooled men" often revealed important lessons, and that the urgency of the situation was such that Americans could not afford to wait for experts to weigh in with their opinions on the subject. "[W]e are," he wrote, "even now, breaking up the floor and wainscoting and doors and window frames of our dwelling, for fuel to warm our bodies and seeth our pottage, and the world cannot afford to wait until the slow and sure progress of exact science has taught it a better economy" (52). In the fifty years following its publication, however, *Man and Nature* probably had its greatest impact among professional foresters. As professionals began to have more influence on the way American forests were utilized, *Man and Nature* was one of the most frequently used source materials favoring forest conservation. The American Association for the Advancement of Science drew heavily upon the book in its memorial on forests prepared for Congress in 1873, which eventually resulted in the creation of the national forest system in 1891 (Lowenthal 268). Nearly every U.S. Forestry Commissioner until well into the 1900s cited *Man and Nature* as the preeminent work in the field, although Gifford Pinchot, the leading forester of the early twentieth century, qualified his praise of this "epoch making" work by saying that it had served mainly to draw the attention of "better-prepared observers" to the issues discussed, but probably had little influence otherwise (xvi–xvii).

Pinchot's qualified praise of Marsh's work may have partly reflected a professional's bias against the work of an amateur, but it probably had more to do with the fact that Marsh's concept of managed forests was taken to its extreme by Pinchot. Although *Man and Nature* is in large measure a utilitarian treatise, unlike Pinchot, Marsh did not feel that the *only* value of wilderness was economic. In calling for the creation of a forest preserve to protect the Adirondacks, for instance, Marsh emphasized the economic over the poetic value of preservation only because he believed that economic issues had previously been neglected by other advocates of forest protection with more romantic notions of why the forests should be preserved. Still, he made it clear that there were valid reasons besides the economic ones:

I believe that the motive of the [Adirondack Park] suggestion has originated rather in poetical than in economical views of the subject. Both these classes of considerations have a real worth. It is desirable that some large and easily accessible region of American soil should remain, as far as possible, in its primitive condition, at once a museum for the instruction of the student, a garden for the recreation of the lover of nature, and an asylum where indigenous tree, and humble plant that loves the shade, and fish and fowl and four-footed beast, may dwell and perpetuate their kind, in the enjoyment of such imperfect protection as the laws of a people jealous of restraint can afford them. (203–204)

As Donald Worster has pointed out, Pinchot saw the goal of forestry simply as maximizing the productivity of the forests, not as preserving "the more complicated biological matrix in which the trees grew" (269), whereas Marsh, certainly concerned with maintaining forest resources, saw this as but one factor necessitating protection of an immensely complex and interconnected system.

In his biography of John Muir, Michael P. Cohen writes: "Men like Marsh . . . pointed the way to a program of 'Resource Conservation and Development.' Perhaps this was a form of ecological consciousness, but it ran in a shallower stream than Muir's" (164). In large measure this is true; however, it seems needlessly to downplay Marsh's contribution in favor of that of Muir, whose response to nature is more in tune with that of many modern environmentalists. The early political successes of the conservation movement were due in no small part to the "resource conservation" arguments Marsh developed in *Man and Nature*, and it was these arguments that overcame both the entrenched economic interests and the wasteful practices of generations of Americans. The changing of American attitudes toward nature and conservation has taken place incrementally. Although Marsh maintained a faith that the American people and their government would finally act to reverse the course of environmental degradation upon which they had embarked, he was cognizant of the difficulties involved in changing wasteful policies and practices that had endured since the colonial era. Perhaps drawing from his experience in Vermont where the legislature had declined to act upon his report on the propagation of fish, Marsh acknowledged the difficulty of persuading a government to invest in a project where the benefits were subtle and long-term. He stressed, nonetheless, that "there can be no doubt that it would be sound economy in the end" (202). It would take years after the publication of *Man and Nature* for conservationists to effect meaningful and widespread change in public policy, which took place most notably during the progressive era of Theodore Roosevelt. Yet, Marsh firmly believed that, once the American people were educated as to the immense benefits of woodlands and the adverse effects of deforestation, they would realize that forest preservation and extension, while costly, "are among the most obvious of the duties which this age owes to those that are to come after it" (279).

Notes

1. Marsh possessed a dry wit, but unfortunately it was displayed only infrequently in *Man and Nature*, and then usually buried in the voluminous footnotes to the text. Although he realized that some of his statements about the railroads and "the present rebellion" in the south were not germane to his thesis, he rationalized their inclusion by stating that, "it is hard to 'get the floor' in the world's great debating society, and when a speaker who has anything to say once finds access to the public ear, he must make the most of his opportunity without inquiring too nicely whether his observations are 'in order'" (51). In one cutting aside, he refers to a southern scientist as someone "whose scientific reputation, though fallen, has not quite sunk to the level of his patriotism" (369).

2. Marsh discussed in some detail the harm caused when farmers killed birds that they mistakenly believed were eating their crops, only to find that in the absence of the birds real destruction was caused by the now unchecked insect population: "Very many of those [birds] generally supposed to consume large quantities of the seeds of cultivated plants really feed almost exclusively upon insects, and frequent the wheatfields, not for the sake of the grain, but for the eggs, larvae, and fly of the multiplied tribes of insect life which are so destructive to the harvests" (80).

3. While in Maine, Thoreau had noted the prevailing lack of respect for public lands, writing, "Much timber has been stolen from the public lands. (Pray, what kind of forestwarden is the Public itself?) I heard of one man who,

G. P. Marsh and the Harmonies of Nature

having discovered some particularly fine trees just within the boundaries of the public lands, and not daring to employ an accomplice, cut them down, and by means of block and tackle, without cattle, tumbled them into a stream, and so succeeded in getting off with them without the least assistance. Surely, stealing pine-trees in this way is not so mean as robbing hen-roosts" (*The Maine Woods* 145).

Works Cited

Brooks, Paul. *Speaking for Nature*. Boston: Houghton Mifflin, 1980.
Cohen, Michael P. *The Pathless Way: John Muir and American Wilderness*. Madison: University of Wisconsin Press, 1984.
Huth, Hans. *Nature and the American*. Berkeley: University of California Press, 1957.
Leopold, Aldo. *A Sand County Almanac*. 1949. New York: Ballantine Books, 1970.
Lincklaen, John. *Travels in the Years 1791 and 1792, Pennsylvania, New York, Vermont*. 1793. In *Vermont: A Chronology and Documentary Handbook*, 77–86. Dobbs Ferry, N.Y.: Ocean Publications, 1979.
Lipke, William C. "Changing Images of the Vermont Landscape." In *Vermont Landscape Images 1776–1976*, William C. Lipke and Philip N. Grime, eds., 33–48. Burlington, Vt.: Robert Hull Fleming Museum, 1976.
Lowenthal, David. *George Perkins Marsh: Versatile Vermonter*. New York: Columbia University Press, 1958.
Marsh, George Perkins. *Man and Nature, or, Physical Geography as Modified by Human Action*. 1864. Cambridge: The Belknap Press of Harvard University Press, 1965.
Mitchell, Lee Clark. *Witnesses to a Vanishing America: The Nineteenth-Century Response*. Princeton: Princeton University Press, 1981.
Mumford, Lewis. *The Brown Decades: A Study of the Arts in America, 1865–1895*. New York: Harcourt, Brace and Company, 1931.
Nash, Roderick. *Wilderness and the American Mind*. 1967. New Haven: Yale University Press, 1982.
Pinchot, Gifford. *Breaking New Ground*. 1947. Seattle: University of Washington Press, 1972.
Thoreau, Henry David. *The Maine Woods*. 1864. New York: Harper & Row, 1987.
Tichi, Cecelia. *New World, New Earth: Environmental Reform in American Literature from the Puritans through Whitman*. New Haven: Yale University Press, 1979.
Worster, Donald. *Nature's Economy: A History of Ecological Ideas*. Cambridge: Cambridge University Press, 1977.

CHAPTER 4

As the Angels Have Departed: John Burroughs and the Religion of Nature

> *His fear of Nature has given place to love. Man never loved as he does now. He has withdrawn his gaze from heaven and fixed it upon the earth. As his interest in other worlds has diminished, his interest in this has increased. As the angels have departed, the children have come in.*
> —John Burroughs, *Time and Change*, 1912

In the August 1871 issue of the *Atlantic Monthly* there appeared an unsigned review by the magazine's editor, William Dean Howells, of a book entitled *Wake-Robin*, a collection of nature essays written by John Burroughs, thirty-three-year-old clerk at the United States Treasury Department. "It is in every way an uncommon book that [Burroughs] has given us," wrote Howells, "fresh, wholesome, sweet, and full of a gentle and thoughtful spirit; a beautiful book" (254). Over the next fifty years, John Burroughs would evolve from the transcendentalist-inspired birdwatcher of *Wake-Robin* to a literary proselytizer for nature second only to John Muir among his literary contemporaries; and in at least one area—book sales—Burroughs outstripped even Muir as a popular chronicler of natural history. At the

John Burroughs and the Religion of Nature 69

time of his death in 1921 over one and a half million copies of Burroughs's books were in print. His immensely popular essays portrayed nature appreciation as a spiritual pursuit on a par with mainstream religion, and helped dissipate lingering American antipathy toward nature. Incorporating the work of Darwin and Whitman with that of the transcendentalists, he created a unique blend of natural history and spirituality that attracted both critical acclaim and a popular audience for nature and nature writing surpassing anything to this point. It was also a crucial step toward the development of a political constitutency for conservation.

Burroughs's lengthy literary career made him a contemporary of Emerson and Thoreau on one side and of Aldo Leopold on the other. When *Wake-Robin* was published in 1871, there was still an American frontier in the west (the battle at Little Big Horn wouldn't take place for another five years), and there were still no national parks or conservation organizations. When Burroughs died in 1921, a headline in the *New York Times* read "Burroughs Began His Study When Buffalo Still Overran the West" (March 30, 1921, 12:3), a notable fact in a modern nation in danger of being overrun by the millions of Model T's produced by Burroughs's friend and admirer Henry Ford. By 1878, Burroughs's reputation as a nature writer was already so well established that he wrote in his journal: "My writing has brought me more fame and money than I ever dared hope" (75). His biographer, Perry Westbrook, asserts that "it is not an exaggeration to say that [Burroughs] enjoyed an esteem and respect comparable to Emerson's" (50). His essays appeared in countless periodicals and were collected in nearly two dozen books. He was honored with numerous awards, including a gold medal from the American Academy of Arts and Letters in 1916 and honorary degrees from Yale, Colgate, and the University of Georgia.[1] Particularly in his later years, Burroughs's Catskill retreats of Slabsides and Woodchuck Lodge became popular pilgrimages for students of nature and admirers of the "Sage of Slabsides," and his visitors included Henry Ford, Thomas Edison, Harvey Firestone, John Muir, and Theodore Roosevelt.

Since his death, Burroughs's literary reputation has faded considerably, and his contributions to environmental reform have largely been minimized or even overlooked entirely. His most recent biographer, Edward Renehan, claims that his essays "encouraged a garden-style nature appreciation" that ignored the industrial excesses of his era (4), and states flatly that Burroughs's name cannot be included among the ranks of environmental activists. Bill McKibben, who is generally appreciative of Burroughs's work, concedes that it sometimes seems "old-fashioned" (32), perhaps even "cloying or overdone" (33). Such criticism is understandable, as Burroughs was more a product of his times than were Thoreau and Muir, the two nature writers to whom he is most often compared. Yet, unlike some other popular nature writers of his era, such as William Long or Ernest Thompson Seton, Burroughs's work has a depth that merits reappraisal of his literary contribution. A close reading of his essays reveals that he was a writer who not only rejected the dominant anthropocentric worldview and displayed a sympathy for and understanding of nature and its intricate relationships, but who was, in his own way, as important in gathering popular support for the conservation movement as any other writer of his era. While Burroughs is no longer generally considered to be the literary equal of Thoreau or Muir, he was probably more successful than they at instilling a love of nature in his readers—although, like Thoreau, he had little interest in the political aspects of protecting nature.

Burroughs was often compared to Thoreau (a critical tendency that continues to plague many nature writers today), and during his lifetime he had a literary reputation that exceeded that of Thoreau. Burroughs was sensitive about the comparison, however, as is clearly evident from the number of times he addressed the issue in his journal: "My current is as strong in my own channel as Thoreau's in his. Thoreau preaches and teaches always. I never preach or teach. I must have a pure result" (*HBJ* 74). Burroughs was uncomfortable with the comparison even where his work was judged to be the superior, as in a November 1910 article in the *Atlantic* by Dallas Lore Sharp. Referring to the laudatory article, Burroughs wrote to a friend, "Why compare *me* to the disadvantage of Thoreau? Thoreau is my master in many ways—much nearer the stars than I am—less human, maybe, but more divine—more heroic" (Barrus, *LL* 2:147). Burroughs occasionally criticized what he saw as Thoreau's antisocial tendencies, but he also claimed to envy Thoreau's professed coolness to human company, feeling that he was himself sometimes too deferential and accommodating. Thoreau, said Burroughs with a mixture of admiration and censure, probably got "more pleasure to open his door to a woodchuck than to a man" (*IS* 269).

It was this human quality in Burroughs's essays, his attempts to get to the meaning behind the fact, that invested them with much of their charm, and, according to Burroughs, made natural history essays interesting for a general audience (*FS* 194). John Burroughs the inquisitive farmer plays a prominent role in the essays by John Burroughs the nature writer, and the latter is at his best when relating anecdotes of farm life and his experiences with the birds and wildlife that frequented his Catskill farm. His descriptions of outdoor life in *Wake-Robin* and *Locusts and Wild Honey*, epitomize the appeal and strength of his work. His narratives of wilderness camping trips are remarkably free of the romanticism that mars much of this type of writing,[2] as he includes the inopportune rainstorms, obscurely marked trails resulting in miles of extra hiking, and the myriad insects that plague visitors to the deep woods. Partly because these details are not glossed over, when an epiphanic moment does occur it is portrayed in its proper context and has a greater impact than it might otherwise have, as in this episode during a camping trip plagued by rain and insects:

> About ten o'clock, as we stood around our campfire, we were startled by a brief but striking display of the aurora borealis. My imagination had already been excited by talk of legends and of weird shapes and appearances, and when, on looking up toward the sky, I saw those pale phantasmal waves of light chasing each other across the little opening above our heads, and at first sight seeming barely to clear the treetops, I was as vividly impressed as if I had caught a glimpse of a veritable spirit of the Neversink. The sky shook and trembled like a great white curtain. (*LWH* 120)

Burroughs's method of turning nature into literature was a deceptively simple one, as he described it in "Straight Seeing and Thinking": "Truth of seeing and truth of feeling are the main requisite: add truth of style and the thing is done" (*LT* 123).

Although Burroughs's main strength as a writer was his power of observation and description, his reflections on the spiritual importance of nature give his work the kind of depth that is absent from the work of many other popular nature writers of his day. Burroughs's linking of nature with spirituality was originally inspired by Emerson, whose essays provided the young Burroughs with an alternative to the fundamentalist religion of his father, Chauncey Burroughs. Burroughs described his father as a basically good man albeit "bigoted and intolerant in his religious and political views" (Barrus, *OFJB* 56). His father's strong religious beliefs remained something of a puzzle to John,[3] but Chauncey Burroughs didn't compel his son to join his church; and John never did, although he did experience a youthful near-conversion to the Methodist faith his father abhorred. Burroughs would later disavow this experience, saying, "I don't believe in these emotional conversions under strong excitement. It's just as exactly as a man does a thing when he's drunk" (Johnston 283). Emerson's essays provided Burroughs with a viable substitute for the personally unfulfilling religious faith of his father. As Edward Renehan writes, Emerson's work became for Burroughs "a philosophy of daring and inspiring affirmation: a revolutionary, natural theology that was the solvent of encrusted forms and traditions" (46). Probably even more important was Emerson's admonition to turn to nature in order to learn about oneself, although it would take some time before Burroughs applied this stricture to his writing. Burroughs's early literary efforts were slavishly derivative of Emerson—so much so that when Burroughs submitted "Expression," a transcendentalist-inspired essay on understanding the universe, to the *Atlantic Monthly* in 1860, James Russell Lowell, the magazine's editor, refused to print the article until he verified that Burroughs had not plagiarized an obscure piece from Emerson (Barrus, *LL* 1:52).

This experience was partly responsible for the turn toward nature writing that Burroughs took thereafter. In order to counterbalance Emerson's influence on his style and subject matter Burroughs began to devote his writings to outdoor themes: "The woods, the soil, the waters, helped to draw out the pungent Emersonian flavor and restore me to my proper atmosphere. But to this day I am aware that a suggestion of Emerson's manner often crops out in my writings" (*IS* 268). In 1863 Burroughs met Emerson after a lecture the latter gave at West Point, and he met with him again several times in later years. He never developed a close friendship with the New Englander, however, as he did with the other great literary influence on his work, Walt Whitman. Still, this did not preclude him from referring to Emerson as "my spiritual father in the strictest sense" and feeling acutely remorseful for his failure to travel to Concord the day Emerson was buried.[4] Despite his effort to put some literary distance between himself and Emerson, Burroughs always acknowledged the great influence Emerson's writings had on him—an influence he shared with nearly all the important nature writers of his era. Long after Burroughs had moved on to other fields of study, traces of transcendentalist thought often cropped up in his work, and he maintained a continuing interest in Emerson throughout his career, writing that Emerson was such an important figure that "attention cannot be directed to him too often" (*TLH* 1).

Although Burroughs's spiritual approach to nature study was, at least initially, grounded in transcendentalism, there is a critical difference in the way Burroughs and Emerson see nature. Burroughs argued that Emerson's anthropocentric philosophy led inevitably to a "subjective and humanized Nature, a Nature of [Emerson's] own creation" (*AU* 22). Burroughs's view of nature, influenced by Darwinian theory, was essentially biocentric. Although Emerson's essays had freed Burroughs from the need to define the natural world by using traditional religious dogma, it was Darwin's work that provided him with a unified concept to replace what myth and religious creeds neatly accounted for—the place of humankind in the world. In *The Origin of Species* (1859) and *The Descent of Man* (1871), Darwin posed an alternative to the biblical version of creation—and if Darwin's theories were correct and humankind shared a common heritage with the rest of the organic world, then a fundamental reappraisal of man's place in the world was necessary. Burroughs first wrote of Darwin in an essay entitled "Analogy," which appeared in *Knickerbocker's Magazine* in December 1862. In that essay, he stated that "Darwin's hypothesis of the derivation of species is in keeping with the unity we everywhere discern."[5] In 1883, Burroughs began to study Darwin in earnest, and after finishing *The Descent of Man* he wrote in his journal that "I have no more doubt of its main conclusions than I have of my own existence" (*HBJ* 98).

The greatest influence on Burroughs was that of Walt Whitman, who served as something of a spiritual complement to Darwin. Whitman was attuned to nature's spiritual side without contradicting what Burroughs saw as its scientific reality.[6] Whitman's influence on Burroughs was so considerable that Burroughs's biographer and literary executor, Clara Barrus, stated that "outside of Nature herself" Whitman had the greatest effect on Burroughs (Barrus, *LL* 2:133), and recounted their relationship at length in *Whitman and Burroughs, Comrades* (1931). The two became close friends while they worked in Washington during the Civil War, and Burroughs's first book, *Notes on Walt Whitman as Poet and Person* (1867), was about (and extensively revised by) Whitman. He would write about the poet many more times over the course of his career, including another full-length work published in 1896, *Whitman: A Study*. Burroughs praised Whitman's use of nature in his poems, particularly the manner in which it was depicted: "He did not humanize nature or read himself into it; he did not adorn it as a divinity; he did not see through it as a veil to spiritual realities beyond, as Emerson so often does. . . . Yet he made more of nature than any other poet has done" (*AU* 322–23). Burroughs believed that Whitman's poetry was so vast and universal that he could reconcile seemingly incompatible concepts, incorporating the best of modern knowledge and humanism with the ancient yearning for religion, writing in *Field and Study*: "He is prophetic and creative, while he is Darwinian and democratic" (230). Whitman's work also reinforced Burroughs's belief that there was a spiritual component to the love of nature. Whitman, says Burroughs, has "his face to earth and not to heaven; he finds the miraculous, the spiritual, in the things about him, and gods and goddesses in the men and women he meets" (*W* 293).

Where natural history was concerned, however, the roles of master and pupil were exchanged, with Burroughs contributing a great deal to Whitman's use of nature in his poetry. In a letter to Myron Benton dated January 9, 1864, Burroughs mentions that he had told Whitman about a camping trip to the Adirondacks he and Benton had taken, and that Whitman had enjoyed the story so much that he had said he "should make a 'leaf of grass' about it" (Barrus, *LL* 1:110). In the same letter, Burroughs says he told Whitman about other country experiences, which the poet "relished hugely." In September 1865, Burroughs wrote to Benton that Whitman "is deeply interested in what I tell him of the Hermit Thrush, and says that he has used largely the information I have given him in one of his principal poems" (Barrus, *WBC* 24). The principal poem of which Burroughs spoke was "When Lilacs Last in the Dooryard Bloom'd," Whitman's elegy for Lincoln, in which the thrush appears as one of the poem's primary symbols.

Although Burroughs never used the word "ecology" (which wasn't used in its modern sense until 1866) in his writing, issues of interconnectedness and man's place in the universe are recurring themes in his nature essays and are often linked to his ideas about God and nature. He addresses the themes of science, nature, and religion at length in books such as *The Light of Day* (1900), *Time and Change* (1912), and *Under the Apple-Trees* (1916).[7] Burroughs's religious opinions appear to have been essentially a mixture of Enlightenment and transcendentalist philosophy, with a generous mixture of Darwin, Whitman, and pantheism added in. Like Darwin himself, Burroughs didn't discount the notion of a supreme being who had originally created the universe (the Enlightenment image of a "great watchmaker" surfaces occasionally in his writing), but after the world had been set in motion, Burroughs felt there was little evidence to indicate that God was intimately involved in human affairs:

Ah me! as soon as we make God out to be a person who interferes in the events of the world, into what strait we are forced! We are forced to conclude that either He is not omnipotent, or else that He is a monster of cruelty. . . . No, He is not that kind of a God. The only way He interferes or takes a hand is through the eternal laws which He has established. (*IS* 248)

This concept of a noninterventionist God is repeated numerous times throughout Burroughs's works and is closely linked to his valorization of nature love as a type of religious experience. Our passion for nature, he wrote, is entwined with our religious instinct (*R* 224); however, the modern influence of science has worked to inject human reason into the equation: "the old myths mean little to us. We accept nature as we find it, and do not crave the intervention of a God that sits behind and is superior to it. The self activity of the cosmos suffices" (*AU* 121).

Burroughs made it clear that he believed some kind of religion was a necessary component of spiritual growth, and this is where the love of nature could fill the vacuum left by the discrediting of the old creeds. "Religion," in Burroughs's definition, meant more than simply embracing a particular creed or dogma, but was rather "a feeling of awe and reverence inspired by the contemplation of this wonderful and mysterious universe, something to lift a man above purely selfish and material ends, and open his soul to influences from the highest heavens of thought" (107). While traditional religions could achieve this, Burroughs considered them to be deficient because they were frequently at odds with reason, and exchanged the real world for a phantom one.[8] Because he was such a popular writer, Burroughs's stance on nature and religion took on a political significance that had a direct impact on environmental discourse. Burroughs may well have been the first American nature writer to state publicly that "the personal, or *anthropocentric*, point of view must be abandoned" (*LD* 53). Reversing Emerson's statement in "The Method of Nature" that, "In the divine order, intellect is primary: nature, secondary" (*CW* 123), Burroughs declares that "Nature is first and man last" (*LD* 205), a theme he explored further in a number of later essays:

Man has from the earliest period believed himself of divine origin. . . . He has spurned the clod with his foot; he has denied all kinship with bird and beast around him, and looked to the heavens above for the sources of his life. And then unpitying science comes along and tells him that he is under the same law as the life he treads underfoot, and that law is adequate to transform the worm into the man. (*TC* 178)[9]

In *Accepting the Universe*, the final collection of essays published before his death, Burroughs asserts that Christianity had not only separated the once coterminous concepts of God and Nature, but had also divorced humankind from nature:

Under the influence of Christianity man has taken himself out of the category of natural things, both in his origin and in his destiny. Such a gulf separates him from all other creatures, and his mastery over them is so complete that he looks upon himself as exceptional, and as belonging to a different order. (20-21)

For Burroughs, like Thoreau, the discounting of the natural world in favor of the supernatural was one of the chief defects of Christianity:

"Our religion is at fault, our saints have betrayed us, our theologians have blackened and defaced our earthly temple, and swapped it off for cloud mansions in the land of Nowhere" (53).

Amid the "decay of creeds," Burroughs believed that the love of nature had a high spiritual value not inconsistent either with our sense of reason or with the part of ourselves that yearns for a higher meaning to life. In a passage that virtually serves as a creed for pantheists, Burroughs writes:

This [love of nature] has saved many persons in this world—saved them from mammon-worship, and from the frivolity and insincerity of the crowd. . . . It has made them contented and at home wherever they are in nature—in the house not made with hands. This house is their church, and the rocks and the hills are the altars, and the creed is written in the flowers of the field and in the sands of the shore. . . . Every walk to the woods is a religious rite, every bath in the stream is a saving ordinance. . . . It is not an insurance policy underwritten by a bishop or a priest; it is not even a faith; it is a love, an enthusiasm, a consecration to natural truth. (AU 116–17)

Burroughs was cognizant that his construct of Nature/God would be considered a denial of God's existence by many adherents of traditional religions (190), and his other comments on conventional religion do little to soften this impression. While he conceded that religious creeds and myths were attempts to "clothe the spirit against the cosmic chill" (256), and as such they would continue to appeal to some, he felt that did not validate them. It is time, he asserted, to exchange the old myths for a system that is consistent with human reason. Christianity might be a "workable hypothesis," he writes, but its philosophy, based on the myth of the fall, was "irrational and puerile," particularly when considered from the perspective of evolutionary theory, which indicated that the "fall" was actually upward (263).

Burroughs had some concern that his essays on religion would prove offensive to the large audience he had attracted through his natural history essays. In a letter to Myron Benton in 1886, he included a copy of his essay "Science and Theology" and asked Benton to "tell me if you think it would produce a commotion among my readers in this country, if published here. *Is it offensive?*" (Barrus, LL 1:281). However, Burroughs's comments on religion appear to have had little if any chilling effect on his growing literary reputation, which is somewhat surprising considering how radical his ideas must have seemed to many of his readers. After publication of *The Light of Day* in 1900, he did receive a few inquiries regarding his religious views from readers, but it appears there was little in the way of outright hostility.[10] There were some isolated complaints that he had too much Nature in his books and not enough God, to which Burroughs responded that this "seems to me like complaining there is too much about the daylight and not enough about the sun" (*FS* 241).

It was not for his philosophy that most readers came to Burroughs, however, but for his natural history, particularly his essays on birdwatching. Burroughs's ornithological fame spread to the point where, on a visit to Scotland, he met a young man with whom he conversed about the native birdlife. Unaware of who the American bird enthusiast was, the young Scot quoted Burroughs to Burroughs, and was surprised to learn that he was talking to the same expert he was citing (Barrus, *LL* 1:240). The extinction of the passenger pigeon, a species Burroughs had seen migrating in enormous numbers through the Hudson River valley in his youth, appears to have been the first instance in which he realized that humankind was capable of exterminating entire species and was in the process of doing so. In *Birds and Poets* (1877) Burroughs wrote: "I am always at home when I see the passenger pigeon. Few spectacles please me more than to see clouds of these birds sweeping across the sky. . . . They come in such multitudes, they people the whole air; they cover townships, and make the solitary places gay as with a festival" (89). In his next book, *Locusts and Wild Honey* (1879), he observed that the large numbers of these birds seemed inexplicable, as they laid only two eggs per season and were the common prey of all, "millions of them meeting a murderous death every year"—and yet, they seemed to thrive (156). In the 1895 edition of *Locusts and Wild Honey*, he added a footnote to this passage stating simply that the pigeon "now seems on the verge of extinction." By 1919, when *Field and Study* was published, the extirpation of the passenger pigeon was complete. Burroughs recalled the last great migration he had seen, and commented bitterly; "The pigeons never came back. Death and destruction, in the shape of the greed and cupidity of man, were on their trail" (14).

Burroughs's own ornithological practices reflected a growing awareness of the effect man's influence could have on bird populations. His early study of the birds included collecting bird specimens, nests, and eggs, and he defended this practice as a necessary part of ornithology.[11] In his first book, *Wake-Robin*, he told of shooting a warbler to identify it, justifying this action by claiming, "no sure and rapid progress can be made in the study [of birds] without taking life, without procuring specimens" (49–50), an assertion with which Thoreau for one would have taken issue. In his next book his attitude toward shooting specimens begins to change somewhat, as he notes that a "good shot with the eye" can also satisfy (*WS* 186). By the latter part of his career, Burroughs's passion for collecting had cooled considerably. In later essays, such as "Bird Enemies," "Birds' Eggs," and "Tragedies of the Nests," which appeared in *The Century* magazine as well as in his books, Burroughs attacked "so-called 'collectors,' men who plunder nests and murder their owners in the name of science" (*SS* 225). He describes with revulsion the work of one of these "human weasels" who worked his way through an orchard "leaving, as he believed, not one nest behind him"—the usually gentle Burroughs then warned grimly, "He had better not be caught working his way through my orchard" (226). Burroughs continued his diatribe against collectors in "Bird's Eggs," where he admonished his readers to "Admire the bird's egg and leave it in its nest" (*R* 67).

This was the image of John Burroughs—nature lover and sage—that took hold in the public's consciousness and inspired numerous bird lovers and incipient conservationists.[12] Biographer Clara Barrus claimed that he had outgrown killing things by midlife, except for "woodchucks and other 'varmints,'" which were always considered fair game to the farmer (*LL* 1:79). In "The Ways of Sportsmen," Burroughs writes, "a man in the woods, with a gun in his hand, is no longer a man,—he is a brute. The devil is in the gun to make brutes of us all" (*R* 305–306). This attitude, however, did not keep him from forming a close friendship with the nation's most famous hunter, Theodore Roosevelt. Unlike John Muir, there is no indication that Burroughs ever called Roosevelt to task for his hunting; in fact, during his trip with Roosevelt to the Yellowstone Park in 1903, he defended Roosevelt's hunting. In turn, Roosevelt dedicated his *Outdoor Pastimes of an American Hunter* to "Oom John," saying; "Your writings appeal to all who care for the life of the woods and fields. . . . It is a good thing for our people that you have lived, and surely no man can wish to have more said of him."

Given his love of nature one might have expected Burroughs to have been more vigorously engaged in public battles on behalf of conservation, but he tended to shy away from such controversies. He did serve as vice president of the Audubon Society, and often spoke at Arbor Day gatherings and similar occasions, but, with the exception of his lobbying efforts in support of the McLane Bird Protection Bill in 1913, he declined to play a significant role in the political battles over issues such as wilderness preservation. He looked on approvingly as Roosevelt expanded the national parks system and his friend John Muir led the fight to save the Sequoia groves of the Sierras, but he seemed to be temperamentally unable to enter the fray. At times Burroughs's reluctance to throw his considerable prestige into the political debate over conservation puzzled, and even irritated, those who felt he should do so. Rather than contributing directly to the movement his writings had helped to inspire, Burroughs chose instead to invest his energies in such subjects as the nature-faker controversy and Bergsonian philosophy, issues that now seem almost quaint. Despite a long stint in Washington as a government clerk, Burroughs generally had little interest in politics, writing in his journal in 1901 that "I think but rarely of politics and the great blundering world" (*HBJ* 226).

Burroughs's lack of political involvement in environmental reform issues may in part be attributed to his attachment to the Catskill farmland of his ancestors. Unlike Muir—who "needed a continent in which to roam"—Burroughs traveled infrequently and had limited firsthand knowledge of the dramatic changes taking place in the American landscape. As Paul Brooks writes, for Burroughs, "travel was an interruption, accepted with reluctance" (13). Ironically, the home-loving Burroughs died on a trainride back to New York from the west coast, but his last words were, fittingly, "Are we home yet?" (*New York Times*, March 30, 1921, 12:3). Save for a few isolated instances, such as the disappearance of the passenger pigeon, Burroughs didn't feel the passing of the wilderness as keenly as Muir did, because his connection with the land was a far more localized, pastoral one than Muir's. Most importantly, Burroughs's rural farmland was not in immediate danger of being forever changed, as Muir's beloved Sierra wilderness was.

What Burroughs did accomplish for the conservation movement was to attract a mass audience for nature, and for literature about nature—an audience that was unprecedented in the United States, and that was crucial to the development of political support for conservation programs and legislation. Without popular support that could be translated into votes, conservationism was little more than an intellectual exercise with scant effect on public policy. In 1887 Burroughs's influence became even more widespread, as his publisher, Houghton Mifflin, brought out a collection of his essays for use in the secondary schools. Clara Barrus claims that this was essentially the beginning of the nature study movement in America (*LL* 1:285). While this point may be debatable, there is little doubt that it contributed substantially both to Burroughs's fame and to instilling a reverence for nature in several generations of American school children. Lawrence Buell points out that Burroughs's popularity also had an indirect effect on Thoreau's, as Houghton Mifflin had so much success with his nature essays that they used his name to help promote Thoreau's work (35). Unfortunately for Burroughs's modern literary reputation, however, both his unassertive prose style and his pastoral, nostalgic vision of America strike many as old-fashioned and irrelevant. Unlike Thoreau's and Muir's, his essays don't have the kind of rhetorical bite that resounds as though they were written in response to today's environmental controversies.

Still, as Edward Renehan writes in his biography of Burroughs, "there is one important thing that redeems Burroughs. In essay after essay, he tried to instill a new, modern element of faith into the faithless decades of the Gilded Age" (4). This faith, the love of nature, is the part of Burroughs's writing that has perhaps the most appeal and relevance for modern readers. Bill McKibben rightly points out that while Burroughs's sentiments regarding the love of nature might not seem revolutionary, "we are still groping toward them" (32). Perhaps most important, the impact that Burroughs's writing had in instilling a love for nature among the generation that would wage—and win—the first great battles over conservation is inestimable. Clara Barrus aptly assessed Burroughs's contribution to conservationist discourse in her rebuttal to the criticism of a more politically active friend: "He failed to appreciate that the forte of John o' Birds did not lie in fighting for anything; and that what he might have done in a militant way for the birds, was but a drop in the bucket to what he did do, in his own way, for more than fifty years" (*LL* 2:313). Burroughs himself was aware that others faulted his political detachment (as they did Thoreau's), but he felt his writing supported the cause, albeit in an indirect way: "I was never a fighter. I fear that at times I may have been a shirker, but I have shirked one thing, or one duty, that I might the more heartily give myself to another" (Barrus, *LL* 2:312).

Notes

1. He was also the subject for a number of paintings and busts. On April 12, 1918, C. S. Pietro's bust of Burroughs, "The Seer," was presented to the city of Toledo, Ohio, and Burroughs was coaxed into attending the unveiling: "The day has arrived. A great crowd. 20,000 school children pass in review before me, bringing flowers. Over one and one-half hours in passing. I stand there on the steps as smiling as a basket of chips. Then I greet the teachers inside the [Art] Museum. Pretty tired at night. All is vanity and vexation of spirit" (Barrus, *LL* 2:270–71).

2. A notable example of this is William Murray's *Adventures in the Wilderness, or, Camp Life in the Adirondacks* (1869), which lured thousands of tourists unprepared for the rigors of outdoor life to the Adirondacks. In addition to the physical hardship—downplayed by Murray in favor of the idyllic scenery—Murray failed to mention the large areas ravaged by lumber and mining operations, and he was roundly condemned in print by many disappointed visitors (Keller 127).

3. Burroughs relates an incident that illustrates just how strong those beliefs were: "I remember when a mere lad hearing him pray in the hog-pen. It was a time of unusual religious excitement for him, no doubt; I heard, and ran away, knowing it was not for me to hear" (Barrus, *OFJB* 56).

4. "April 30, Sunday [1882] Today Emerson is to be buried, and I am restless and full of self-reproach because I did not go to Concord. I should have been there. Emerson was my spiritual father in the strictest sense. It seems as though I owe nearly all, or whatever I am, to him. . . . I fell in with him just in time. His words were like the sunlight to my pale and tender genius which had fed on Johnson and Addison and poor Whipple.

"I must devote the day to meditating on Emerson, the greatest and most typical of all New-Englanders" (*HBJ* 87–88).

5. The essay appears in a revised form in *Literary Values* (1902).

6. Perry Westbrook, in his critical biography of Burroughs, emphasizes Whitman's influence, and asserts that their friendship was mutually beneficial, as Burroughs was one of Whitman's most ardent and persuasive early supporters.

7. Regarding *The Light of Day*, Norman Foerster asserts that, "On its negative side the book is so belated that the writer is really knocking down a man of straw—the authority of revealed religion had already been destroyed when Burroughs entered the lists" (288). However, while this may have been true to a large extent among the scientific community, it is far less clear that this was the case with the general audience Burroughs was actually addressing.

8. The necessity of turning our faces "to earth and not to heaven" is one that Burroughs explored in his studies of Whitman, and was closely tied to Whitman's role as a champion of democracy. Burroughs wrote that, "Carried out in practice this democratic religion will not beget priests, or churches, or creeds, or rituals, but a life cheerful and full on all sides, helpful, loving, unworldly, tolerant, open-souled, temperate, fearless, free, and contemplating with pleasure, rather than alarm, 'the exquisite transition of death'" (*W* 293).

9. The worm striving to be man is an allusion to Emerson's epigraph to "Nature": "A subtle chain of countless rings / The next unto the farthest brings; / The eye reads omens where it goes, / And speaks all languages the rose; / And, striving to be man, the worm / Mounts through all the spires of form."

10. Clara Barrus, Burroughs's biographer and literary executor, gives one instance where an irate reader challenged his views on religion and evolution ("that scientific lie") and accused Burroughs of having a "swelled head and intellectual pride." Burroughs responded by facetiously asking his correspondent, "I should like to ask you how you think God made man—with his clothes on? his finger nails pared? and his hair cut? Did he say to man 'Come forth!' and at one stroke original man stepped out from a bank of clay just as we know him today?" (Barrus, *LL* 2:401).

11. The mania for collecting, particularly where it coincided with nature study, was a peculiar characteristic of the late nineteenth century, and large glass cases filled with bird specimens were a common ornament in many households. In reference to one such case created by Burroughs, Walt Whitman said that it was a "poem."

12. The contradictions present in some of Burroughs's essays appear to have gone largely unnoticed or were just uncritically accepted by his admirers. For instance, in *Wake-Robin* he calls the hawk an "audacious marauder" and

then shoots one (101). The unintended irony in some of these passages can be rather amusing, such as an incident described in *Riverby* where Burroughs wonders what enemy the weasel has that would cause him to seek refuge in such a deep and complicated burrow—while he is digging mightily in an attempt to unearth the creature (124).

Works Cited

Barrus, Clara, ed. *The Life and Letters of John Burroughs* (2 vols.), Boston: Houghton Mifflin, 1925.
———. *Our Friend John Burroughs* (cited as *OFJB*). Boston: Houghton Mifflin, 1914.
———. *Whitman and Burroughs, Comrades* (cited as *WBC*). Boston: Houghton Mifflin, 1931.
Brooks, Paul. *Speaking for Nature*. Boston: Houghton Mifflin, 1980.
Buell, Lawrence. "Henry Thoreau Enters the American Canon." In *New Essays on Walden*, Robert F. Sayre, ed., 23–52. New York: Cambridge University Press, 1992.
Burroughs, John. *Accepting the Universe* (cited as *AU*). New York: Russell & Russell, 1920.
———. *Birds and Poets*. Boston: Houghton Mifflin, 1875.
———. *Field and Study* (cited as *FS*). New York: Russell & Russell, 1919.
———. *The Heart of Burroughs's Journals* (cited as *HBJ*). Clara Barrus, ed. Boston: Houghton Mifflin, 1928.
———. *Indoor Studies* (cited as *IS*). Boston: Houghton Mifflin, 1889.
———. *Leaf and Tendril*. Boston: Houghton Mifflin, 1908.
———. *Light of Day* (cited as *LD*). Boston: Houghton Mifflin, 1900.
———. *Literary Values*. Boston: Houghton Mifflin, 1902.
———. *Locusts and Wild Honey* (cited as *LWH*). Boston: Houghton Mifflin, 1879.
———. *The Last Harvest* (cited as *LH*). New York: Russell & Russell, 1922.
———. *Notes on Walt Whitman as Poet and Person*. New York: J. S. Redfield, 1871. Reprinted by University Microfilms, Ann Arbor, Mich., 1964.
———. *Riverby* (cited as *R*). Boston: Houghton Mifflin, 1894.
———. *Signs and Season* (cited as *SS*). Boston: Houghton Mifflin, 1886.
———. *Time and Change* (cited as *TC*). Boston: Houghton Mifflin, 1912.
———. *Under the Apple Trees*. New York: Russell & Russell, 1916.
———. *Wake-Robin*. Boston: Houghton Mifflin, 1871.
———. *Whitman: A Study* (cited as *W*). Boston: Houghton Mifflin, 1896.
———. *Winter Sunshine* (cited as *WS*). Boston: Houghton Mifflin, 1875.
Emerson, Ralph Waldo. *The Collected Works of Ralph Waldo Emerson*. Vol. 1. Cambridge: The Belknap Press of Harvard University Press, 1971.
Foerster, Norman. *Nature in American Literature: Studies in the Modern View of Nature*. New York: Russell & Russell, 1923.
Johnston, Clifford, ed. *John Burroughs Talks*. Boston: Houghton Mifflin, 1922.
McKibben, Bill. "The Call of the Not So Wild." *The New York Review of Books*, May 14, 1992, 32–33.
Murray, William H. H. *Adventures in the Adirondacks*. 1869. Syracuse: Syracuse University Press, 1989.

Renehan, Edward. *John Burroughs: An American Naturalist*. Chelsea Green Publishing Company: Post Mills, Vt., 1992.
Roosevelt, Theodore. *The Wilderness Hunter/Outdoor Pastimes of an American Hunter I*. New York: Charles Scribner's Sons, 1926. Vol. 2 of *The Works of Theodore Roosevelt: National Edition* (20 vols.).
Westbrook, Perry. *John Burroughs*. New York: Twayne Publishers Inc., 1974.
Whitman, Walt. *Leaves of Grass*. 1855. New York: Penguin Books, 1959.

CHAPTER 5

The God of the Mountains: The Rhetoric and Religion of John Muir

> *These temple destroyers, devotees of raging commercialism, seem to have a perfect contempt for Nature, and, instead of lifting their eyes to the God of the mountains, lift them to the Almighty Dollar.*
> —John Muir, *The Yosemite*, 1913

In 1869, thirty-one-year-old John Muir took a job as a shepherd in the foothills of California's Sierra mountain range, getting his first glimpse of the region where he would spend much of the next several years. He later described his first reaction to the mountains in terms that reflect a type of religious ecstasy:

Through a meadow opening in the pine woods I see snowy peaks about the headwaters of the Merced above Yosemite. . . . How consuming strong the invitation they extend! Shall I be allowed to go to them? Night and day I'll pray that I may, but it seems too good to be true. Some worthy will go, able for the Godful work, yet as far as I can I must drift about these love-monument mountains, glad to be a servant of servants in so holy a wilderness. (FS 16)

For the next several years, Muir lived in the Yosemite valley, using it as a base for extensive travels in the mountains of California. Much of his later writing would describe this region, combining natural history with a pantheistic spiritualism that makes transcendentalism seem pale in comparison. "Heaven knows that John Baptist was not more eager to get all his fellow sinners into the Jordan than I to baptize all of mine in the beauty of God's mountains," Muir wrote in his journal in 1871 (*JOM* 86), and through his nature essays many readers were indeed "baptized" as Muir had hoped. As John Tallmadge points out, "many elements of his natural theology (particularly those consonant with an ecological paradigm) have influenced generations of nature writers and, through them, the shape of environmental thinking today" (78).

It is hard to overstate the importance of John Muir's contribution to the wilderness preservation movement in America. Muir's work galvanized the incipient conservation movement. Roderick Nash writes in *Wilderness and the American Mind*: "As a publicizer of the American wilderness Muir had no equal" (123), and Norman Foerster rates Muir and John Burroughs as the outstanding nature writers since Thoreau (264). In his essays Muir built a case for preserving the wilderness combining the spiritual qualities of Emerson and Thoreau with the hard-edged practicality of George Perkins Marsh. But, unlike Thoreau and Burroughs, the other two great nature writers of the nineteenth century, Muir's involvement with environmental reform went beyond writing books and magazine articles, to include direct political action. Following his successful battle to have the Yosemite valley brought under the protection of the federal government in 1905, Muir wrote to Robert Underwood Johnson: "I am now an experienced lobbyist; my political education is complete. Have attended Legislature, made speeches, explained, exhorted, persuaded every mother's son of the legislators, newspaper reporters, and everybody else who would listen to me" (Bade 1:356). In 1892 Muir founded the Sierra Club, which played a key role in preserving numerous wilderness areas during his lifetime and continues to spread Muir's message of environmental protection through political activism to this day.

Muir's accomplishments as a lobbyist for the wilderness are all the more surprising given the fact that his writing shows him to be every bit as contrarian and misanthropic as Thoreau is often perceived to be, and he is even more emphatic in his statements on the religious value of nature. In addition, Muir's break with the anthropocentric view of man's place in nature is even more clearly stated than Thoreau's. Whereas Thoreau, as John Burroughs said, "would rather open his door to a woodchuck than to a man," Muir goes a step further, writing, "if a war of races should occur between the wild beasts and Lord Man, I would be tempted to sympathize with the bears" (*SBY* 343). However unconventional Muir's biocentric views of nature might have been, he was so adept rhetorically that he was able to persuade enormous numbers of Americans that wilderness preservation was in the best interest of the nation.

Muir's passion for the wilderness and his effectiveness as a polemicist both have their roots in the religious upbringing he rejected. Ironically, the biblical imagery that was so much a part of Muir's upbringing would later be used to convince his readers of the spiritual value of nature. As John Elder writes, "Biblical language is the vehicle with which Muir expresses his discovery of spirit throughout nature" (385). Muir described his religious upbringing as a boy in Scotland at some length in *The Story of My Boyhood and Youth* (1912), recalling that "father made me learn so many Bible verses every day that by the time I was eleven years of age I had about three-fourths of the Old Testament and all of the New by heart and by sore flesh" (27). In 1849 Muir's father, Daniel, moved the family from Dunbar, Scotland, to the Wisconsin prairie, where the religious indoctrination of his children continued. Daniel Muir used the bonfires in which they burned the brush cleared from the land as warning lessons, "comparing their heat with that of hell, and the branches with bad boys" (63). However, the palliative effect of the wilderness had already begun to insinuate itself into eleven-year-old John's consciousness: "those terrible fire lessons quickly faded away in the blithe wilderness air; for no fire can be hotter than the heavenly fire of faith and hope that burns in every healthy boy's heart." By the time Muir was in his teens he had begun to bridle under his father's authoritarian rule, and it was over his awakening interest in science and inventing that he and his father argued most frequently. Muir persuaded his father to allow him to read some books on higher mathematics and history, but his father balked at Dick's *Christian Philosopher*, wary of the word "philosophy" in the title. When John defended the book, holding that there were kinds of useful philosophy we could not do without, his father countered that "the Bible is the only book human beings can possibly require throughout all the journey from heaven to earth" (*SBY* 194). Next, using a rhetorical tactic he would later employ in his arguments for forest preservation, Muir approached the issue from a purely practical stance, pointing out that without a "little helpful science" there wouldn't be people able to make the spectacles his father needed to read the Bible in the first place: "But he still objected to my reading that book, called me a contumacious quibbler too fond of disputation, and ordered me to return it to the accommodating owner. I managed, however, to read it later" (*SBY* 194).

In *The Story of My Boyhood and Youth*, Muir relates another series of debates with his father that usually concerned their differing interpretations of the Bible. On one occasion, after his father had taken to a vegetable and graham flour diet, which he tried to force on the rest of the family, John convinced him that the Bible did not advocate such a diet. He referred to the scriptural passage where the prophet Elijah had been fed on flesh provided by ravens sent from God: "The Bible being the sole rule, father at once acknowledged that he was mistaken. The Lord never would have sent flesh to Elijah by the ravens if graham bread were better" (195). These debates with his father over religion and learning taught Muir many of the basic rhetorical skills that served him so well in his later life. From an early age he saw how effectively the use of logic and the proper authority—in his father's case, the Bible—could sway a seemingly intractable opponent. While Muir's personal reasons for wilderness preservation were spiritual, he was remarkably adept at couching his arguments in utilitarian terms when he knew that these would have a greater effect on his intended audience.

Muir's accounts of his early reaction to his father's religious dogmatism provide a revealing glimpse into his movement away from Christian fundamentalism and toward a spirituality more in harmony with what he saw in nature. While *The Story of My Boyhood and Youth* was written at some remove from the experiences described and may characterize these youthful experiences in a manner calculated to reflect Muir's later philosophy, there is little doubt that Daniel Muir's fundamentalism was inverted—either purposefully or subconsciously—by John Muir's religion of nature. One of the prime tenets in nature as Muir later presented it in his writing was the unity of all living things, a unity that ran contrary to the dichotomy between man and the rest of the natural world that his father's religion espoused. In a powerful passage, Muir tells of his father's inordinate fondness for "almost every sort of church meeting, especially revival meetings" (*SBY* 88), and how that fondness led to the death of Muir's favorite horse, Nob. His father had driven the horse to Portage, Wisconsin, and back, a twenty-four-mile roundtrip, on a hot summer day. That night the horse took sick, and Muir connected its suffering directly to his father's overweening thirst for religion:

It was a hot, hard, sultry day's work, and [Nob] had evidently been overdriven in order to get home in time for one of those meetings. I shall never forget how tired and wilted she looked that evening when I unhitched her; how she drooped in her stall, too tired to eat or even to lie down. Next morning it was plain that her lungs were inflamed; all the dreadful symptoms were just the same as my own when I had pneumonia. (*SBY* 88)

It is significant that when Muir describes his own bout with pneumonia later in the book, it is again linked to his father's fanaticism for religion and hard work: "Only once was I allowed to leave the harvest-field—when I was stricken down with pneumonia. I lay gasping for weeks, but the Scotch are hard to kill and I pulled through. No physician was called, for father was an enthusiast, and always said and believed that God and hard work were by far the best doctors" (*SBY* 88). Muir saw this insensitivity to the suffering of others as a direct outgrowth of religious fanaticism, characteristic not only of his father. He tells of an instance where a "half-witted" neighbor attempted to commit suicide by drowning, which was seen by the man's brother "simply as a crime calculated to bring harm to religion" rather than as a "terrible proof of despair" (*SBY* 172).

Muir's empathy with the horse and its affliction is the most striking aspect of his description of its "weary suffering and loneliness of the shadow of death" (*SBY* 88) and is crucial to fully understanding his rejection of an anthropocentric approach to nature. Nob's death taught Muir a lesson about man's kinship with other creatures that was far more enduring than his father's stories of hellfire:

She was the most faithful, intelligent, playful, affectionate, human-like horse I ever knew, and she won all our hearts. Of the many advantages of farm life for boys, one of the greatest is the gaining a real knowledge of animals as fellow-mortals, learning to respect them and love them, and even to win some of their love. Thus godlike sympathy grows and thrives and spreads far beyond the teachings of churches and schools, where too often the mean, blinding, loveless doctrine is taught that animals have neither mind nor soul, have no rights that we are bound to respect, and were made only for man, to be petted, spoiled, slaughtered, or enslaved. (*SBY* 89)

Muir's biocentric belief in the unity among all living creatures is reiterated many times in his writing. In *The Story of My Boyhood and Youth* he relates an incident where some of the family's hogs narrowly escaped being killed for food by some passing Indians, writing that he would never forget the "fear in the eyes of that old mother and those little pigs." The pigs' terror, writes Muir, "was as unmistakable and deadly a fear as I ever saw expressed by any human eye, and corroborates in no uncertain way the oneness of all of us" (73). Muir's most popular book among contemporary audiences, *Stickeen* (1909), also dealt with this theme. Stickeen was a small dog Muir befriended while he was exploring glacier fields in Alaska, and with whom Muir had narrowly escaped death during one episode on the glaciers. In his journal notes regarding the incident, Muir wrote that Stickeen "was the herald of a new gospel," having once again demonstrated to him that "human love and animal love, hope and fear, are essentially the same, derived from the same source, and fall all alike like sunshine" (*JOM* 277). He noted that whereas "Indian dogs go to the Happy Hunting Grounds with their masters," in "civilized religions" all mortal creatures except man are selfishly shut out (*JOM* 277).[1]

Muir believed that the hardening effect of religion extended as well to the way religious extremists like his father behaved toward other people. In *The Story of My Boyhood and Youth*, he describes, with some bitterness, the unnecessarily harsh drudgery of life on the Muir farm. On one occasion, his father had him dig a well down ninety feet through sandstone, chipping away laboriously at the rock with mason's chisels. One day when he was lowered by his father into the well he was nearly overcome by the carbonic acid gas that had settled at the bottom of the pit during the night; he was pulled out of the well by his father and brother just before losing consciousness. The resentment he felt toward his father regarding this incident is apparent even though he is describing it many years later: "Constant dripping wears away stone. So does constant chipping, while at the same time wearing away the chipper. Father never spent an hour in that well" (186).

The philosophical estrangement between Muir and his father continued even after John had left the family farm to embark on the travels that would take him to the University of Wisconsin at Madison,[2] and ultimately to California. Several years after Muir left the farm his father would himself leave his home and family for missionary work in Canada. In 1872, while living in Yosemite, John learned that his father had plans to sell the family farm, so he wrote to his brother David asking him to intervene on behalf of their mother and sisters. It is obvious from the tone of his letter that John had as dim a view of his father's spiritual calling as his father had of his, telling David, "I expected a morbid and semi-fanatical outbreak of this kind as soon as I heard of his breaking free of the wholesome cares of the farm" (Bade 1:24). Fortunately for the welfare of the family, David was successful in persuading his father to continue his evangelical work while maintaining the home in Wisconsin. For his part, Daniel Muir never became reconciled to his son's work. He wrote a letter to John in Yosemite, after reading his son's published account of a night spent on Mount Shasta during a violent storm. In it he urged John to renounce his mountain studies and embrace Christ and God's work, saying "the best and soonest way of getting quit of the writing and publishing of your book is to burn it, and then it will do no more harm either to you or others" (Bade 1:22).

Following a course of studies at the University of Wisconsin at Madison that concluded in 1863, Muir held a series of odd jobs in

Canada and the Midwest, continuing the inquiries into botany that he had begun while at the university. After nearly losing his eyesight in a factory accident, Muir began a walking tour from Indianapolis to Florida that commenced in September 1867 and concluded in January 1868, just prior to his voyage to California. The journal notes from this tour would later be edited by his literary executor, William F. Bade, and published as *A Thousand-Mile Walk to the Gulf* in 1916. Muir's writing from this period shows a continued movement away from the fundamentalism of his father, but more important, it shows that even at this early age Muir was beginning to see nature itself as the source of his own religious inspiration.

The sense of liberation from both his father's lingering influence and the drudgery of the factory work he had just escaped is reflected by the "address" Muir inscribed on the inside cover of his journal notebook—"John Muir, Earth-planet, Universe." Given the enormous distance, both geographical and philosophical, that Muir would travel during the next several years, it seems appropriate that he described the outset of his pilgrimage in terms similar to those often used in traditional Christian narratives of spiritual discovery:

As soon as I got out into heaven's light [following the eye injury], I started on another long excursion, making haste with all my heart to store my mind with the Lord's beauty, and thus be ready for any fate light or dark. And it was from this time that my long continuous wanderings may be said to have fairly commenced. I bade adieu to mechanical inventions, determined to devote the rest of my life to the study of the inventions of God. (SBY 236-37)

Given his familiarity with biblical imagery, it is natural that Muir frequently lapsed into the literary style he was most familiar with, that of the King James Bible. He depicts his travels in terms reminiscent of the flight of the Jews from Egypt, calling to mind the "long continuous wanderings" that signified a release from bondage and journey toward the promised land—in Muir's case, not Israel, but Yosemite. Later, during his stay in Yosemite, Muir would frequently describe the valley in religious terms, writing his brother David, for instance, to tell him that, although he had not been to church since leaving home, "this glorious valley might well be called a church" (Bade 1:209).

In *A Thousand-Mile Walk to the Gulf* Muir considers at some length the false dichotomy he believed had been set up between man and nature. He conjectures that there is a "human" element to the plants he studies that is not yet understood: "How little we know as yet of the life of plants—their hopes and fears, pains and enjoyments!" (SBY 262). In a passage resembling Thoreau's "Chesuncook" in its pantheism,[3]

Muir writes: "They tell us that plants are perishable, soulless creatures, that only man is immortal, etc.; but this, I think, is something we know very nearly nothing about. Anyhow, this palm was indescribably impressive and told me grander things than I ever got from human priest" (*SBY* 319). In passages such as this, Muir is obviously critical of religious teaching that assumes only man among all life on earth has a divine element, a soul—but his questioning of religious dogma here has implications beyond the theological. Muir's biocentric theology subtly turned the fundamentalist Christian teachings of his father into a spiritual ecology that is one of the precursors of deep ecology.[4]

While Muir doesn't challenge the old tenet that the chief end of man is to serve God, he does set out to counter an implicit corollary of that belief: that the chief end of nature is to serve man. Muir criticizes the anthropocentric way in which man tends to look at nature—judging everything by how it serves or pleases him—and that the common belief that "repellant" creatures such as snakes and alligators were created not by God but by Satan, "thus accounting for their all-consuming appetite and ugliness" (*SBY* 324). Although we may not understand the place in nature of such animals, he writes, God does, providing and caring for them despite our "dismal irreverence": "Though alligators, snakes, etc., naturally repel us, they are not mysterious evils. They dwell happily in these flowery wilds, are part of God's family, unfallen, undepraved, and cared for with the same species of tenderness and love as is bestowed on angels in heaven or saints on earth" (*SBY* 324). As on his father's farm, where Muir identified with the stricken horse Nob, here he professes a kinship even with those animals commonly considered to be dangerous, such as the alligator, to which he says, "may you long enjoy your lilies and rushes, and be blessed now and then with a mouthful of terror-stricken man by way of dainty" (*SBY* 325).

As Muir's journey progressed southward, he grew more emphatic in his statements against the presumption that the world was made especially for "Lord Man"—"a presumption not supported by all the facts" (*SBY* 355). Muir took issue not with the notion of a creator, but with the way God was portrayed by humans who endowed the Creator with human sympathies in order to better justify their own actions:

A numerous class of men are painfully astonished whenever they find anything, living or dead, in all God's universe, which they cannot eat or render in some way what they call useful to themselves. They have precise dogmatic insight of the intentions of the Creator, and it is hardly possible to be guilty of irreverence in speaking of *their* God any more than of heathen idols. He is regarded as a civilized law-abiding gentleman in favor either of a republican form of government or of a limited monarchy; believes in the literature and language of England; is a warm supporter of the English constitution and Sunday schools and missionary societies; and is as surely a manufactured article as any puppet of a half-penny theater. (*SBY* 355)

With such a conceited notion of the Creator, Muir concluded, it is "not surprising that erroneous views should be entertained of the creation" (*SBY* 355).

Muir's assaults on the anthropocentric view of nature in what was intended to remain a private journal (William Bade published it as *A Thousand-Mile Walk to the Gulf* several years after Muir's death) occasionally go far beyond anything he would later write for a general audience—an audience he hoped to convert to the cause of wilderness preservation. In a passage that leaves little doubt of where his sympathies lie he writes:

> Let a Christian hunter go to the Lord's woods and kill his well-kept beasts, or wild Indians, and it is well; but let an enterprising specimen of these proper, predestined victims go to houses and fields and kill the most worthless person of the vertical godlike killers,—oh! that is horribly unorthodox, and on the part of the Indians atrocious murder! Well, I have precious little sympathy for the selfish propriety of civilized man, and if a war of races should occur between the wild beasts and Lord Man I would be tempted to sympathize with the bears. (*SBY* 343)

While Muir may not have flaunted such misanthropic sentiments in his persuasive writing, it appears he did little to conceal them in conversation. In a letter to a mutual friend, John Burroughs complained of Muir's fondness for argument, writing: "conversation with him is a sparring-match with gads. He likes to get in the first cut and follow it up. . . . Yet see how tender Muir assumes to be towards the animals! Yet he likes to walk over the flesh of his fellow man with spurs in his soles" (Barrus, *LL* 2:185). Like Thoreau, who spoke for wildness because "there are enough champions of civilization" (*EP* 205), Muir felt compelled to do the same, and was often just as *"extra-vagant"* as Thoreau in his pronouncements on the value of wilderness.

Like John Burroughs, however, Muir believed that the love of nature provided one with a measure of protection against what Burroughs termed "the cosmic chill." One of the most vivid sections of *A Thousand-Mile Walk to the Gulf* deals with a period when Muir was stranded near Savannah, Georgia, waiting for his brother to wire him some money so that he could resume his journey. Without adequate funds to rent a room, he decided to sleep in a nearby cemetery, reasoning: "There no superstitious prowling mischief-maker dares venture for fear of haunting ghosts, while for me there will be God's rest and

The Rhetoric and Religion of John Muir

peace" (305). There, in the "weird and beautiful abode of the dead," Muir stayed for a week, until the expected package from his brother arrived at the post office. Muir's narrative of his stay at the Bonaventure cemetery is notable both for its beautiful descriptive passages and for its considerations of death, of which Muir writes: "On no subject are our ideas more warped and pitiable" (302). We are, he continues, commonly taught that death is something to be feared and dreaded, something that is—like the cemetery itself—"an ill-omened place, haunted by imaginary glooms and ghosts of every degree" (302). The fear of death associated with this attitude is sharply contrasted with the peace of mind that comes from being attuned to nature:

But let children walk with Nature, let them see the beautiful blendings and communions of death and life, their joyous inseparable unity, as taught in the woods and meadows, plains and mountains and streams of our blessed star, and they will learn that death is stingless indeed, and as beautiful as life, and that the grave has no victory, for it never fights. All is divine harmony. (303)

This theme of the "divine harmony" is one that Muir returns to again and again throughout his writings, as are many of the other opinions expressed in *A Thousand-Mile Walk to the Gulf*, such as the connection between God and nature, the unity of man and nature, and the refutation of the notion that nature was created solely for man's use. Even before Muir reached California, then, he had essentially formulated a new system of beliefs to replace the rejected fundamentalism of his father, and lacked only a specific focus, a "temple," for his religious fervor and energies.

After a bout with malaria dashed a rather ill-conceived plan to follow his thousand-mile walk and a brief trip to Cuba with a trek up the Amazon, Muir sailed to San Francisco in early 1868. In 1869 he spent the summer herding sheep in the Sierras, an experience he described in *My First Summer in the Sierras*, published many years later in 1911. He was so taken with the Yosemite region that he spent the next several years there, working and studying the geology and plant life of the area. Muir's Yosemite experiences reinforced his religious views of nature; so much so in fact, that John Tallmadge suggests that *My First Summer in the Sierras* is a conversion story similar to those of St. Paul and St. Augustine (62). Muir's studies of the region were also important in that they led him to publish (at the urging of friends) his first essays. These early essays were primarily an outgrowth of his studies of Yosemite's residual glaciers and of his theory that the valley had been created by glacial action rather than by a cataclysmic upheavel of the earth's surface. This theory had involved Muir in his first public debate, with Professor Josiah D. Whitney, the California state geologist and author of the *Yosemite Guide-Book*. In a series of letters published in the New York *Tribune* and articles that appeared in *The Overland Monthly*, *Scribner's*, and *Harper's*, Muir set forth his theories of glacial action in the Yosemite.[5] He challenged the findings of Whitney and his survey team (which included the noted geologist and mountaineer Clarence King), asserting that the reason the survey team and other visitors to the region had not come to the same conclusion as he had was because "The labors of the State Geological Survey in this connection amounted to a slight reconnaissance. . . . [T]he common tourist, ascending the range only as far as Yosemite Valley, sees no portion of the true Alps containing the glaciers excepting a few peak clusters in the distance (*Harper's Magazine*, November 1875, 776). Whitney and King attacked Muir's findings. King referred to Muir as an "ambitious amateur," although he did express the unintentionally prescient (if hostilely phrased) hope that Muir would "divert his evident enthusiastic love of nature into a channel, if there is one, in which his attainments will save him from hopeless floundering" (Farquhar 163). As Francis Farquhar writes in his *History of the Sierra Nevada*, later researchers proved Muir's interpretation of the geological origins of Yosemite to be substantially correct, and he is now recognized as having discovered living glaciers in the Sierra Nevada (163).[6]

In addition to pursuing his glacier studies, Muir traveled throughout the mountains of California and read extensively, finding the transcendentalists to be particularly useful. Muir had first been exposed to the writings of Emerson and Thoreau during his studies at the University of Wisconsin in the early 1860s, although it appears that he didn't read *Walden* or the majority of Emerson's essays and poems until his Yosemite period (Wolfe 79). The philosophy of nature espoused by the transcendentalists buttressed Muir's own beliefs, and he frequently cited Emerson and Thoreau with approval.[7] His regard for the two men was such that when he visited their gravesites at Concord's Sleepy Hollow Cemetery in 1893 he wrote to his wife: "I did not imagine I would be so moved at sight of the resting-places of these grand men as I found I was, and I could not help thinking how glad I would be to feel sure that I would also rest there" (Bade 2:267). One of the great highlights of Muir's years in the Yosemite was a visit by Emerson in 1871, which he described in *Our National Parks*. Muir proposed to take Emerson on a camping trip amid the big trees of the Mariposa Grove, but Emerson's Bostonian companions stymied the plan, arguing that he might take cold (146). Muir spent several days in Yosemite with the philosopher nevertheless, and when Emerson left to return to Boston,

Muir recalled, "I felt lonely, so sure had I been that Emerson of all men would be the quickest to see the mountains and to sing them" (148–49). Although the two men were never to meet again, they maintained a correspondence, and Emerson invited Muir to visit him in Concord. In Emerson's journal was a list of "My Men"; the final name added to the list was that of John Muir.

While traces of Emerson's influence frequently surface in Muir's writing, and Muir certainly had a high regard for both the man and his work, Muir's taste for argument was such that he still found the need to argue with him—in the marginalia of Emerson's books, if not personally. As Stephen Fox notes, Emerson's appreciation of nature had its base in abstract metaphysics, while Muir felt that nature itself was the basis for a proper philosophical approach. Emerson's view of nature was also insufficiently wild for Muir (Fox 83). In his collection of Emerson's essays, Muir carried on a vicarious debate with the great philosopher, disputing various points of natural history and metaphysics:

Emerson: "The squirrel hoards nuts, and the bee gathers honey, without knowing what they do." Muir: *How do we know this.*
Nature "takes no thought for the morrow." *Are not buds and seeds thought for the morrow.*
"It never troubles the sun that some of his rays fall wide and vain into ungrateful space, and only a small part on the reflecting planet." *How do we know that space is ungrateful.*
"The soul that ascends to worship the great God is plain and true; has no rose-color." *Why not? God's sky has rose color and so has his flower.*
"The beauty of nature must always seem unreal and mocking, until the landscape has human figures." *God is in it.*
"The trees are imperfect men, and seem to bemoan their imprisonment, rooted in the ground." *No.* (Quoted in Fox, 6)

Yosemite itself was a far more important element in the development of Muir's philosophy than metaphysics, from whatever source, could ever be. As Muir's early journal entries, such as those that form *A Thousand-Mile Walk to the Gulf*, indicate, he had already broken from religious orthodoxy long before he reached California, but Yosemite provided a focal point for both his spirituality and his preservationist rhetoric.

Muir's religious beliefs regarding nature are particularly important because, unlike John Burroughs (whose own religious beliefs have much in common with Muir's), his association of nature with the divine translated directly into political activism in an effort to protect "God's first temples." Muir's religious beliefs have been discussed by a number of commentators, who have reached differing conclusions as to whether his love of nature indicated an implicit rejection of Christianity. His literary executor, Frederick Bade, unconvincingly suggested—perhaps in the interest of protecting Muir's reputation—that Muir had no lasting reaction to his father's fanaticism and exhibited no "lifelong antagonism to religion" (Bade 1:63). Linnie Marsh Wolfe, who wrote the first in-depth biography of Muir, came to a different conclusion, asserting that "John Muir early revolted against religion as he saw it practiced" (38) and that Muir's increasing substitution of the words "Nature" or "Beauty" for "God" or "The Lord" in his manuscripts reveals a pantheistic sense of nature, which she likens to Wordsworth's (267). Perhaps most telling is the religious training Muir gave to his own children: "Very unorthodox he was also in their religious training. God was to be revered not as a person, but as a loving, intelligent spirit creating, permeating, and controlling the universe. They were not taught to pray, except on one occasion Muir told his wife he wished they might know the Lord's Prayer" (Wolfe 232). Later biographers have tended to agree with Wolfe. Stephen Fox suggests that Muir's thousand-mile walk represented a "permanent break from Christianity" (50), and characterizes his later religious orientation as "pantheistic" (70, 80). Michael P. Cohen describes Muir as "pantheistic, ecstatic, and possessed by the cosmic vision" (124), a position that is well supported by Muir's writings.[8]

Muir's writings offer ample evidence that he regarded God and Nature as indivisible concepts and that organized religion tended to be far too narrow-minded to satisfy him: "Imperious bolt upright exclusiveness upon any subject is hateful, but it becomes absolutely hideous and impious in matters of religion, where all men are equally interested" (Bade 1:217). Through nature, people could come into direct communion with the divine, with "the world . . . a church and the mountains altars" (FS 250). Muir's nature love resembled his father's Christian fundamentalism in the way he preached about the spiritual benefits of the wilderness. Not only did Muir proselytize for nature "like a minister preaching the gospel" (FS 147) but he reserved his greatest scorn for those who were either blind to its truths[9] or, even worse, who desecrated the wilderness: "the hills and groves were God's first temples, and the more they are cut down and hewn into cathedrals and churches, the farther off and dimmer seems the Lord himself" (FS 146). With this view of nature, it seems natural Muir felt compelled to take action to preserve the wilderness. "He would write an article, take Teddy Roosevelt camping—*anything* for the cause of natural preservation," writes John Elder (382).

In 1879, ten years after Muir had begun to live in the Yosemite valley, there was still only one national park, Yellowstone, and there were no national forest preserves. Muir's role in creating the system of national parks and forest preserves that exists today was critical, and his essays and books were central in the effort.[10] While his personal motivation in preserving wilderness areas was primarily spiritual, he realized that arguments for preservation based solely on spiritual grounds would find a limited audience. Therefore, he combined utilitarian reasons for preservation with the spiritual, emphasizing such disparate benefits of wilderness as watershed protection and the salubrious effect of the outdoors on one's health. Muir knew that if wilderness preservationists were to prevail over sawmill owners, ranchers, miners, and others with entrenched economic interests they had to persuade the general public that their interests were better served by preserving the wilderness than by allowing vested interests to exploit it unfettered by any governmental restrictions.

The implications of deforestation had already been explored in some depth by Alexander von Humboldt and George Perkins Marsh, among others. Muir was familiar with these works, and in his own writing sought to incorporate their warnings about deforestation in a way that would be accessible and persuasive to a wider audience than those writers had attracted. One of his professors at the University of Wisconsin, Ezra S. Carr, had been strongly influenced by Marsh's work and had passed on this interest to Muir. While Muir was at the University of Wisconsin he also met Increase S. Lapham, the state's foremost conservationist of that period, and was impressed by Lapham's arguments for conservation, many of which were, again, drawn from Marsh's *Man and Nature*. Muir later used some of those arguments in his own writings, making a careful study of *Man and Nature*, as is evident from the large number of marginal notes contained in his own copy of the book (Wolfe 83). In his biography of Muir, Michael P. Cohen acknowledges the influence of Marsh's work on Muir but makes the point that Muir questioned Marsh's anthropocentric rationales for forest preservation when he later formulated his own biocentric ideas (20–21). While there is a great deal of truth to this statement, the importance of Marsh's work, particularly as an authoritative source of conservation theories that backed up Muir's own preservationist ideas and rhetoric, shouldn't be underestimated.

The most challenging task in Muir's effort to create a constituency for wilderness preservation was to persuade an audience—most of whom had never seen the American wilderness—that it was worth protecting. As Bill Devall and George Sessions point out in *Deep Ecology: Living as if Nature Mattered*, modern advocates of environmental reform are often forced to use the language of resource economists in making the case for wilderness protection (3). So was Muir, but despite this impediment he never fell into the trap of confining his arguments to a cost-benefit analysis of wilderness protection. Nor did he rely exclusively on the esthetic or spiritual benefits of environmental protection, which, as Devall and Sessions also point out, can leave environmental reform activists open to charges that they are "sentimental, irrational, or unrealistic" (3). Instead, Muir combined the different rationales for wilderness protection into narratives that so effectively illustrated how important wilderness was that there was no need for him to engage in an extended justification for preserving it. To this end, Muir's descriptive talents were invaluable—indeed, although he sometimes professes himself at a loss for words to describe what he saw (and occasionally overuses adjectives such as "glorious"), his best descriptive essays, such as "A Wind Storm in the Forests" or "The Water-Ouzel," are among the finest in American nature writing. Muir's essays weren't intended to serve as a substitute for the outdoors as much as to entice his readers into experiencing nature for themselves: "But the best words only hint its charms. Come to the mountains and see" (MC 1:235).

Although Muir often wrote about the beauty of the mountains, he also emphasized that preserving the wilderness provided tangible advantages to society. Among these benefits was spiritual and physical relief for those who used nature as an antidote for the industrial age: "Thousands of tired, nerve-shaken, over-civilized people are beginning to find out that going to the mountains is going home; that wildness is a necessity; and that mountain parks and reservations are useful not only as fountains of timber and irrigating rivers, but as fountains of life" (ONP 3). While there is a certain validity to Herbert F. Smith's claim that Muir "found commodity in all its forms a necessary evil at best and a corrupting force at worst (138), he often used commodity as a persuasive tool. Like Marsh, he understood the importance of forests in maintaining watersheds (ONP 350; MC 1:213) and preventing floods (MC 1:287), and he often worked this information into his essays. Muir's first nationally published article on wilderness protection appeared in *Harper's Magazine* in 1877 and dealt with the destruction of the sequoia forests of northern California. In this essay, entitled "God's First Temples: How Shall We Preserve Them?" he called for federal legislation to protect the forests, the first of many such appeals during the next forty years of preservationist work.[11]

The rhetorical technique Muir uses in his wilderness protection essays generally follows a recognizable pattern. For instance, in "The Sequoia and General Grant National Parks," an essay included in *Our*

National Parks, he begins by describing the sequoia groves in superlatives, pointing out some of their most distinctive features and noting the incredible age of the trees, all intended to give a reader a sense of the unique beauty of the park. He then depicts the ravages of a lumber mill situated in the sequoia forest, creating "a sore, sad center of destruction" (323). Leaving the image of the lumber mill lurking darkly in the background, Muir describes the rest of the park. At the close of the essay he returns to the subject of the lumber mills and the danger they pose to the redwood forests (354–55), a danger the perceptive reader may already have divined for himself. By his vivid descriptions of the park, Muir so effectively conveys a sense of the value of the groves that by the end of the "tour" no doubt many of his readers have already anticipated the need for protective legislation as called for in the essay's conclusion.

Muir's line of argument was similar in "The American Forests," one of his best known essays, included in *Our National Parks* (it also appeared in the *Atlantic Monthly* 80:145). He described the idyllic beauty of the American continent when it was first discovered and how the vast forests contained hundreds of useful species of trees. However, rather than wisely using these resources, the early settlers, "claiming Heaven as their guide," proceeded to wage "interminable forest wars; chips flew thick and fast; trees in their beauty fell crashing by millions, smashed to confusion, and the smoke of their burning has been rising to heaven more than two hundred years" (*ONP* 362). After describing the wanton destruction of America's forests, declaring that every civilized country in the world ("Even Japan") did a better job of protecting its forests than the United States, Muir posed the issue of forest protection as one that had reached crisis proportions: "Now it is plain that the forests are not inexhaustible, and that quick measures must be taken if ruin is to be avoided. . . . [L]aws in existence provide neither for the protection of the timber from destruction nor for its use where it is most needed" (*ONP* 367).

Anticipating opposition from lumber interests and others who benefited from the exploitation of the forests, Muir categorized those who would oppose a national policy of forest protection as people who either know nothing about the subject—"That a change from robbery and ruin to a permanent rational policy is urgently needed nobody with the slightest knowledge of American forests will deny" (*ONP* 372) —or "thieves who are wealthy and steal timber by wholesale" (*ONP* 389). Despite the substantial damage already done to the forests, Muir emphasized that it was not too late for the federal government to act, but, he warned, "if the remnant is to be saved at all, it must be saved quickly" (*ONP* 392). Muir concluded his essay with a passionate call to action that exemplified both the power of his rhetoric and the religious fervor that fueled it, describing the trees in terms that show that he empathized with them as living beings, not merely as resources:

> Any fool can destroy trees. They cannot run away; and if they could, they would still be destroyed,—chased and hunted down as long as fun or a dollar could be got out of their bark hides, branching horns, or magnificent bole backbones. . . . Through all the wonderful, eventful centuries since Christ's time—and long before that—God has cared for these trees, saved them from drought, disease, avalanches, and a thousand straining, leveling tempests and floods; but he cannot save them from fools,—only Uncle Sam can do that (*ONP* 393).

While resource management advocates such as Pinchot and Roosevelt might have criticized this depiction of trees as sentimental, its political implications were certainly apparent to them, as was the case with the nature-faker controversy.[12] If people saw trees as something more than lumber—as living creatures in need of protection—many would call for their protection. Roosevelt was far too skilled a politician to ignore such a call even where he might not have agreed with the reasoning behind it, particularly since he was disposed to be on the side of wilderness protection anyway.

At first, Muir and his allies in the wilderness preservation movement were most successful in influencing public policy when they worked in conjunction with conservationists such as Pinchot and Roosevelt. Due in no small part to the efforts of Muir, three national parks—Yosemite, Sequoia, and General Grant—were created in California by act of Congress in 1890. The following year Congress passed the Forest Reserve Act of 1891, which, due to political infighting over the designation of which areas were to be designated as reserves and how that designation would affect timber harvesting, remained a largely symbolic measure for several years. Muir and Pinchot had initially believed their ideas for forest protection to be compatible, but by 1896, when the two men were chosen to serve on a forestry commission appointed by Congress intended to break the deadlock over the Forest Reserve Act, their differing philosophies on forest use had become starkly apparent. While Muir conceded that Pinchot's principles of forest management had some merit, he felt that their primary charge was to determine which areas should be protected from development. Pinchot, on the other hand, believed that the task of the commission was to prepare the forests for managed harvest (Nash 136). The following year, Pinchot recommended that the newly protected forest reserves be opened to grazing and some mining. Muir, who considered sheep to be "hoofed locusts" that would destroy the reserves, confronted Pinchot, asking if he'd been quoted correctly. When Pinchot admitted that he had, Muir declared, "then . . . I don't want anything more to do with you" (Wolfe 275). The term *conservation*, which had previously been construed broadly enough to include both "wise use" advocates and preservationists, was appropriated by Pinchot and his allies, and the extent of the break between Muir and Pinchot soon became apparent when a plan to dam the Hetch-Hetchy valley in order to supply water for San Francisco was proposed in 1900. In a series of articles written for national magazines, Muir extolled the beauty of the valley, comparing it favorably to Yosemite and arguing that San Francisco's need for water could be satisfied without sacrificing one of the country's most beautiful wilderness areas. In "Hetch Hetchy Valley," Muir refuted, one by one, all the arguments for damming the valley, likening them to "those of the devil, devised for the destruction of the first garden" (*Yosemite* 715). Throughout the essay he posed the question starkly as one of good vs. evil, portraying those who would dam the valleys as infidels seeking to destroy one of God's most beautiful creations:

These temple destroyers, devotees of ravaging commercialism, seem to have a perfect contempt for Nature, and instead of lifting their eyes to the God of the mountains, lift them to the Almighty Dollar.
 Dam Hetch Hetchy! As well dam for water-tanks the people's cathedrals and churches, for no holier temple has ever been consecrated by the heart of man. (716)

As in many of his earlier essays, Muir's prose sounds as though it were derived from scripture, and he plays on the dam/damn pun here as he would in other articles on Hetch-Hetchy. In one passage he alludes simultaneously to two biblical sources, drawing an explicit parallel between those in favor of the dam and the money-changers Christ threw out of the temple, and likening the despoliation of the Hetch-Hetchy to the expulsion from Eden: "Thus long ago a few enterprising merchants utilized the Jerusalem temple as a place of business instead of a place of prayer . . . and earlier still, the first forest reservation, including only one tree, was likewise despoiled" (714).

Although Pinchot and the city of San Francisco strongly supported the project, Roosevelt's Secretary of the Interior, Ethan Hitchcock, opposed it. President Roosevelt was quietly supportive of the plan but he refused to overrule Hitchcock, and thus, while the public debate raged, the dam project languished in bureaucratic deadlock. Following the inauguration of William Howard Taft in 1909, proponents of the water project continued their efforts to get the necessary permits for the project while Muir and his allies countered by persuading Taft's new Secretary of the Interior, Richard Ballinger, to delay the project until it could be proven that it was absolutely necessary. Finally, in 1913, a bill was introduced in Congress (supported by Gifford Pinchot) that would grant the Hetch-Hetchy valley to the city of San Francisco for the purpose of constructing a reservoir on the site. In December of 1913—a full thirteen years after the initial proposal to dam the valley—the bill ceding Hetch-Hetchy to San Francisco was signed into law by President Woodrow Wilson.

On December 24, 1914, less than a year later, Muir died of pneumonia. On a personal level, his reputation as an advocate for wilderness was enhanced by the Hetch-Hetchy battle even though he ultimately failed to stop the dam. Ironically, the dramatic rhetoric of his final, losing battle resonated even more powerfully than that of his earlier, more successful efforts to promote the establishment of a system of national parks and forest reserves. But while the loss of Hetch-Hetchy represented a major political defeat for Muir personally and for the preservationists generally, the setback was not without its benefits, as the campaign was a valuable learning experience for the Sierra Club and other preservationists, and served to galvanize public opinion regarding such depredations on the national parks. Perhaps even more important in the long run was the skillful way in which Muir brought spiritual and ethical issues into the debate over environmental reform, and brought the assumptions regarding an anthropocentric view of the world into question. In both a political and a philosophical sense, the roots of the modern environmental movement can be traced directly to Muir and his work. As Muir wrote shortly before his death: "The people are now aroused. Tidings from near and far show that almost every good man and good woman is with us. Therefore be of good cheer, watch, and pray and fight!" (*JOM* 437).

Notes

1. For more on Muir's respect for the rights of animals, see Lisa Mighetto, "John Muir and the Rights of Animals," *Pacific Historian* 29:2–3 (1985): 103–12.

2. Given the deep and abiding differences between Muir and his father it is remarkable that they remained on speaking terms at all. When John left for the university, he asked his father if he would send money should he be in need of any, and his father told him that he wouldn't, and that John should depend entirely on himself: "Good advice, I suppose, but surely needlessly severe for a bashful, home-loving boy who had worked so hard" (*SBY* 209).

3. "It is the living spirit of the tree, not its spirit of turpentine, with which I

sympathize, and which heals my cuts. It is as immortal as I am, and perchance will go to as high a heaven, there to tower above me still" (Thoreau, *The Maine Woods* 165).

4. See Bill Devall, "John Muir as Deep Ecologist." Max Oelschlaeger's *The Idea of Wilderness* 301–306 also discusses Muir's work in the context of the intellectual and philosophical roots of deep ecology. For an in-depth discussion of deep ecology see Bill Devall and George Sessions, *Deep Ecology: Living as if Nature Mattered*.

5. When Professor Joseph LeConte, a geology professor at the University of California, visited Yosemite in 1870, he and his party were accompanied by Muir. After Muir told LeConte of his theories regarding the glacial origins of the valley LeConte wrote in his journal: "[Muir] further believes that the Valley has been wholly formed by causes still in operation in the Sierra—that the Merced Glacier and the Merced River and its branches . . . have done the whole work" (Bade 1:285). LeConte and later visitors such as John Daniel Runkle, president of MIT, urged Muir to write out his theories, which he finally decided to do in late 1871.

6. In her biography of Muir, Linnie Wolfe quotes Francois E. Matthes, author of the *Geological History of the Yosemite Valley* (U.S. Geological Survey, Professional Paper 160, Government Printing Office, 1930), as saying that Muir "was more intimately familiar with the facts . . . and more nearly right in their interpretation than any professional geologist of his time" (187). In his *History of the Sierra Nevada*, Francis Farquhar writes: "Modern geologists agree substantially with John Muir and recognize him as the discoverer of living glaciers in the Sierra Nevada" (163). For more on Muir and the Yosemite glacier controversy see Bart O'Brien, "Earthquakes or Snowflakes," *Pacific Historian* 29:2–3 (1985): 30–41; and Dennis R. Dean, "John Muir and the Origin of the Yosemite Valley," *Annals of Science* 48:5 (Great Britain, 1991): 453–85.

7. Muir sometimes echoes passages from the essays of Emerson and Thoreau, as in this passage from *The Mountains of California*, which recalls Emerson's "transparent eye" section from "Nature": "You are all eye, sifted through and through with light and beauty. Sauntering along the brook that meanders silently through the meadow from the east, special flowers call you back to discriminating consciousness" (1:147). In his journal he writes, "In God's wildness lies the hope of the world," echoing Thoreau's statement on wildness from "Walking" (*JOM* 317), and he quotes both writers a number of times in essays such as "The American Forests."

8. In *Wilderness as Sacred Space*, Linda Graber discusses the concepts of geopiety and sacred space, and links the development of a spiritually oriented wilderness ethic with both personal behavior and political action. To someone who associates nature with religious experience, it is not Nature itself that is being worshipped, but rather "the Power that reveals itself in Nature" (3). Her analysis of the wilderness ethic indicates that the attitudes expressed by Muir in his writings are typical of a geopiety that is essentially pantheistic (5).

9. In "Mormon Lilies" Muir tells of meeting a "grave old Mormon" with whom he'd had some previous discussions on theology: "I shook my big handful of lilies in his face and shouted, 'Here are the true saints, ancient and Latter-Day, enduring forever!' After he had recovered from his astonishment he said, 'They are nice'" (*ST* 134).

10. See, for example, *John Muir and the Sierra Club: The Battle for Yosemite* by

Holway R. Jones; *The Politics of Wilderness Preservation* by Craig W. Allin; and *The Pathless Way: John Muir and American Wilderness* by Michael P. Cohen.

11. The bulk of Muir's preservationist essays were collected in *The Mountains of California* (1893), *Our National Parks* (1901), and *The Yosemite* (1912), although a preservationist ethic is threaded throughout his more autobiographical works such as *The Story of My Boyhood and Youth* (1913) and *My First Summer in the Sierras* (1911).

12. The nature-faker controversy and its political implications for Roosevelt are discussed in more detail in chapter 6.

Works Cited

Allin, Craig W. *The Politics of Wilderness Preservation.* Westport, Conn.: Greenwood Press, 1982.

Bade, William Frederick, ed. *The Life and Letters of John Muir* (2 vols.) Boston: Houghton Mifflin, 1924.

Barrus, Clara, ed. *The Life and Letters of John Burroughs* (cited as *LL*) (2 vols.). Vol. 2. Boston: Houghton Mifflin, 1925.

Cohen, Michael P. *The Pathless Way: John Muir and the American Wilderness.* Madison: University of Wisconsin Press, 1984.

Devall, Bill. "John Muir as Deep Ecologist." *Environmental Review* 6:1 (1982): 63–86.

Devall, Bill and George Sessions. *Deep Ecology: Living as if Nature Mattered.* Salt Lake City: Peregrine Smith Books, 1985.

Elder, John. "John Muir and the Literature of Wilderness." *Massachusetts Review* 22:3 (Summer 1981): 375–86.

Farquhar, Francis. *History of the Sierra Nevada.* Berkeley: University of California Press, 1965.

Foerster, Norman. *Nature in American Literature: Studies in the Modern View of Nature.* New York: Russell & Russell, 1923.

Fox, Stephen. *John Muir and His Legacy.* Boston: Little, Brown and Company, 1981.

Graber, Linda. *Wilderness as Sacred Space.* Washington, D.C.: Association of American Geographers, 1976.

Jones, Holway R. *John Muir and the Sierra Club: The Battle for Yosemite.* San Francisco: Sierra Club Books, 1965.

Muir, John. *John of the Mountains: The Unpublished Journals of John Muir* (cited as *JOM*). Linnie Marsh Wolfe, ed. Madison: The University of Wisconsin Press, 1938.

———. "Living Glaciers of California." *Harper's New Monthly Magazine* (November 1875): 769–76.

———. *The Mountains of California* (cited as *MC*). (2 vols.) Boston; Houghton Mifflin, 1916.

———. *My First Summer in the Sierras* (cited as *FS*). Boston: Houghton Mifflin, 1911.

———. *Our National Parks* (cited as *ONP*). Boston: Houghton Mifflin, 1916.

———. *Steep Trails* (cited as *ST*). Boston: Houghton Mifflin, 1918.

———. *Stickeen.* Boston: Houghton Mifflin, 1909.

———. *The Story of My Boyhood and Youth* (cited as *SBY*) (includes *A Thousand-Mile Walk to the Gulf*). Boston: Houghton Mifflin, 1916.
———. *The Yosemite* (cited as *Y*). 1913. Included in *The Eight Wilderness Discovery Books*. Seattle: The Mountaineer Books, 1992.
Oelschlaeger, Max. *The Idea of Wilderness*. New Haven: Yale University Press, 1991.
Tallmadge, John. "John Muir and the Poetics of Natural Conversion." *North Dakota Quarterly* 59:2 (Spring 1991): 62–79.
Thoreau, Henry David. *The Maine Woods*. 1864. Joseph J. Moldenhauer, ed. Princeton: Princeton University Press, 1972.
———. "Walking." In *Excursions and Poems* (cited as *EP*). Boston: Houghton Mifflin, 1906.
Wolfe, Linnie Marsh. *Son of the Wilderness: The Life of John Muir*. New York: Alfred A. Knopf, 1945.

CHAPTER 6

The Days of Wasteful Plenty Are Over: Theodore Roosevelt and His Environmental Legacies

> Here in the United States we turn our rivers and streams into sewers and dumping-grounds, we pollute the air, we destroy forests, and exterminate fishes, birds, and mammals—not to speak of vulgarizing charming landscapes with hideous advertisements. But at last it looks as if our people were awakening. Many leading men, Americans and Canadians, are doing all they can for the Conservation movement.
> —Theodore Roosevelt, "Our Vanishing Wildlife," *Literary Essays*, 1913

During the weekend following Leon Czolgosz's shooting of President William McKinley in Buffalo on September 6, 1901, McKinley's recovery appeared so certain that Vice President Theodore Roosevelt was told he need not remain in Buffalo. Indeed, McKinley's advisors thought it would be more reassuring to the public if Roosevelt resumed his planned vacation in the Adirondacks (Morris 739). The following week, Roosevelt, his family, and a few other companions left Camp Tahawus, a hunting and fishing resort, for a planned ascent of Mount Marcy, at 5,344 feet the highest peak in New York State. They stayed at Lake Colden the night of September 12, and when the next day's weather proved cold and wet, most of the party, save Roosevelt and a couple of the men, elected to stay in camp. By early afternoon, Roosevelt had reached Marcy's summit and was preparing to return to the camp to rejoin his family when he saw a guide approaching from below; as he would later write in his autobiography: "I felt at once that he had bad news and sure enough, he handed me a telegram saying that the President's condition was much worse and that I must come to Buffalo immediately. . . . That evening I took the oath of office, in the house of Ansley Wilcox, at Buffalo" (339).

This Olympian descent from the mountains to assume the presidency is in keeping both with the symbolism (self-created and otherwise) of Roosevelt and his public career and with the effect he had on the American conservation movement. As Lee Clark Mitchell observes, "To write a history of conservation for our times would be to start with Theodore Roosevelt's friendship with Gifford Pinchot, which led to the most sustained public movement ever on behalf of conservation" (273). Not only did Roosevelt create more national parks, forest reserves, and national monuments than any other president before or since, but he was in large measure responsible for creating the constituency for conservation that made later environmental reforms politically feasible. However, Roosevelt's conservation legacy, both as a chronicler of outdoor life and as president, has become increasingly ambiguous as environmentalism has evolved. A Rooseveltian policy, advocating conservation of resources primarily for their "wise use" in an economic sense, is no longer considered to be progressive at all by many in the environmental movement;[1] and his writings, which emphasized hunting and frontier values, share even less common ground with modern environmentalism. As Donald Worster points out, "in a sense, not much was new in this Progressive conservation; nature was still valued chiefly as a commodity to be used for man's economic success" (262). Seen from this perspective, Roosevelt's descent from the mountain to assume the presidency seems less like that of a giant striding down to rescue the conservation movement and more like a symbolic lowering of principles in order to settle for what was politically feasible.

Both Roosevelt's political philosophy and his writings on outdoor life reflected his firm conviction of the value of the "strenuous life," which served as his personal corollary to "the survival of the fittest." Roosevelt believed that hunting and vigorous outdoor activities helped to build character, and in "The Strenuous Life," a speech delivered in 1899, he argued that national greatness depended upon "the doctrine of the strenuous life, the life of toil and effort, of labor and strife"

(13:319). Roosevelt felt he had overcome poor health during his childhood by applying these principles (20:30-45), and in his historical works, such as *The Winning of the West*, he attributed the growth of the American nation in large part to the virtues developed by the demands of frontier life. As pioneer life gave way to a modern industrial society, Roosevelt was afraid that this would lead to a "softening of fibre" he felt was symptomatic of modern life (2:59). In many ways, *The Winning of the West* anticipated Frederick Jackson Turner's 1893 thesis on the closing of the American frontier; in fact, it was cited by Turner in his work. While in some respects Roosevelt's outdoor antidote for modern life was reminiscent of John Muir's wilderness prescription for "tired, nerve-shaken, over-civilized people," his support for preserving wilderness areas had more to do with nationalism and individual character than with spirituality.

Still, for all the ambiguities inherent in Roosevelt's writing and politics (particularly where the two intersected), it is clear that his espousal of conservation reflected deeply held beliefs regarding the value of nature and outdoor life. The "strenuous life" was an idealization of a taste Roosevelt acquired in boyhood—indeed, Paul Cutright argues that his record as a conservationist was largely an outgrowth of his early interest in natural history (*TMN* xi). As a boy, Roosevelt compiled an extensive collection of specimens and oddities, which he referred to as the "Roosevelt Museum of Natural History" (*TMN* 8), and his early ambition was to be a naturalist like Audubon or Wilson (20:25).[2] It wasn't until after he was well into his undergraduate studies at Harvard that he abandoned his plan to become a scientist. He did continue his field studies, however, and went on extended camping trips in the forests of Maine and the Adirondacks, publishing a monograph of his ornithological observations in the Adirondacks in 1877. It was during these trips that Roosevelt's interest in hunting, which began as an adjunct to his mania for collecting specimens, became an avocation in itself. This aspect of outdoor life would become particularly important to him during his period as a rancher in the Badlands of South Dakota during the 1880s.

Following the deaths of his mother and wife in February of 1884, the grief-stricken Roosevelt spent much of the next several years on his ranches in South Dakota, a period of immense significance in his growth as a writer and conservationist. His time in that region inspired the books on ranchlife and hunting that are among his best literary works and gave him a firsthand look at the fast-disappearing American frontier. Like Francis Parkman, Roosevelt wanted to experience frontier life before it disappeared forever, as he knew it soon would.

While he accepted this change as a necessary part of the progress of civilization, he believed that with the passing of the frontier, America was in danger of losing an important part of its national identity:

> In its present form stock-raising on the plains is doomed, and can hardly outlast the century. The great free ranches, with their barbarous, picturesque, and curiously fascinating surroundings, mark a primitive stage of existence as surely as do the great tracts of primeval forests, and, like the latter, must pass away before the onward march of our people; and we who have felt the charm of the life, and have exulted in its abounding vigor and its bold, restless freedom, will not only regret its passing for our own sakes, but must also feel real sorrow that those who come after us are not to see, as we have seen, what is perhaps the pleasantest, healthiest, and most exciting phase of American existence. (1:292–93)

As a chronicler of life in the west during this period, Roosevelt excelled. His powers of observation were exceptional. He was also a compelling story-teller, and a number of incidents in his ranch life series epitomize what we now see as "the Old West," such as his narrative about capturing outlaws and bringing them to justice (1:379ff.).[3] Contemporary reviewers were generally quite favorably disposed to Roosevelt's ranch life books; a New York Times review of Hunting Trips of a Ranchman predicted that the book "would take a leading position in the literature of the American sportsman" ("New Publications," New York Times, July 13, 1885).

Despite their undeniable merit as literary and historical vignettes of frontier life, Roosevelt's accounts of ranching and hunting in the west contain many passages in which his attitude toward nature has more in common with the nineteenth century than the twentieth. Hunting Trips of a Ranchman (1885), Ranch Life and the Hunting Trail (1888), and The Wilderness Hunter (1893) all present a thoroughly anthropocentric view of the natural world, with little reflection about man's place in nature and the effect of human actions on the land and its creatures. While Roosevelt was aware that the disappearance of the frontier brought with it the end of more than merely a way of life, as evidenced by his comments on the fate of the buffalo,[4] his early writing touched on this aspect only rarely. He chose instead to concentrate on "the game-shooting that forms perhaps the chief of the cattleman's pleasures" (1:25). Even if we don't judge his work by contemporary standards, his descriptions of hunting trips frequently betray an attitude toward wildlife that is symptomatic of his mixed legacy.

Roosevelt's work occasionally exhibits an ethical detachment from his own actions that results in some startling incongruities between his words and his actions and lends a measure of sad and unintentional irony to some of his writing. For example, in one passage from *Hunting Trips of a Ranchman*, he decries the fact that the great elk of the western states "can hold their own and make head against all their brute foes; it is only when pitted against Man the Destroyer that they succumb in the struggle for life." But in the very next sentence he writes, "I have never shot any elk in the immediate neighborhood where my cattle range; but I have had good sport with them in a still wider and more western region" (211). Although Roosevelt frequently emphasizes the fact that most of his hunting was done to obtain meat or a particularly fine trophy, and he often deplores "game-butchery" and unsportsmanlike methods of hunting, his writing abounds with instances where his reason for killing is indefensible even by his own standards. In one incident described in *The Wilderness Hunter* he shoots at an eagle as it soars overhead simply to see if he can hit it: "my bullet grazed his shoulder, and down he came through the air, tumbling over and over. As he struck the ground he threw himself on his back, and fought against his death with the undaunted courage proper to his brave and cruel nature" (58). Despite Roosevelt's derision of game-butchery and bag counts, he occasionally succumbed to the temptation himself, and in *African Game Trails* he itemizes a list of over five hundred animals shot by himself and his son on safari. He then proudly declares that "we did not kill a tenth, nor a hundredth, part of what we might have killed had we been willing" (391).

Although Roosevelt's fondness for hunting has often been disparaged both in his time and in ours, it should be emphasized that he followed the generally accepted game management practices of his time. Like Aldo Leopold, he saw conservation and hunting as complementary, and advocated a "middle ground ... between brutal and senseless slaughter and the unhealthy sentimentalism which would just as surely defeat its own end by bringing about the eventual total extinction of the game" (3:73). Where Roosevelt differs most sharply from modern environmentalists is not over responsible hunting in and of itself, but over issues such as the place of non-game animals in the ecosystem, and policy toward animals threatened with extinction. Like most outdoorsmen of his era, Roosevelt found no fault with the destruction of animals characterized as "varmints," a broadly defined term he applied to species such as the prairie dog (1:160), snakes (1:148), and predators such as wolves and bobcats (1:18).[5] The lack of concern for non-game animals exhibited by Roosevelt reflected a traditional devaluation of those elements of nature not useful to man in some obvious way, a viewpoint that Marsh and Muir sought to rebut in their writing. This prejudice was reflected throughout Roosevelt's writing; even where he called for wilderness preserves, it was for the purpose of protecting game animals and explicitly excluded protection of "noxious species" (3:192). The most pernicious aspect of this prejudice was the unrelenting war against predators that would later be incorporated into the Progressives' conservation program; as Donald Worster writes, "for the first time, the resources of the federal government were brought to bear against the predator. Instead of relying on the varmint-blasting frontiersman, the government itself undertook to eliminate the predator once and for all" (262).

While Roosevelt's attitude toward non-game animals is understandable, his stance on the hunting of buffalo, elk, and other species threatened with extinction is more puzzling. Although he expressed an apparently sincere regret over their decline, this was always tempered by his belief that such occurrences were in the natural course of events and were governed by scientific principles. He describes the status of beaver and elk in the Bad Lands as "close to extinction," but that fails to deter him from hunting them.[6] In *Hunting Trips of a Ranchman*, he notes how the elk had nearly vanished from the area near his ranch, attributing this in part to "their occasional fits of stupid panic, during whose continuance hunters can now and then work greater slaughter in a herd" (208). He was aware that this slaughter meant the elk were "seemingly doomed to total destruction at no distant date" (209), a fate that "can be looked upon only with unmixed regret by every sportsman and lover of nature" (210). However, Roosevelt seems to be mourning the end of elk-hunting more than the end of the elk; when he hears that a few stragglers have been sighted near his ranch, he sets out and kills "probably the last of his race that will ever be found in our neighborhood," with little evident regret (226). Roosevelt characterized the slaughter of the buffalo by white hunters as a "savage war" and called their seemingly inevitable extinction "a veritable tragedy of the animal world" (1:186). The rapidity (and recentness) of their passing was evidenced by the fact that their trails, dried dung, and skeletons were still seen in the western states: "No sight is more common on the plains than that of a bleached buffalo skull; and their countless numbers attest the abundance of the animal at a time not so very long past" (1:186). Despite his awareness of both the scope and barbaric nature of the buffalo's demise, as with the elk Roosevelt was driven to shoot one before they disappeared entirely; the hunting trip is graphically described in *Hunting Trips of a Ranchman* (197).

It is evident that Roosevelt had a certain "survival of the fittest" fatalism about the buffalo that enabled him to rationalize his actions—shooting a "few stragglers"—as inconsequential; the buffalo's doom had been sealed long before he ever picked up a rifle, and was, in a sense, preordained. Roosevelt's comments on the buffalo were also representative of nineteenth-century attitudes toward progress, nature, and the frontier as seen in works such as Francis Parkman's *The Oregon Trail*.[7] Although Roosevelt and Parkman considered the passing of the buffalo and the end of the frontier to be regrettable, in their linear view of history this was all part of the ineluctable progress of American civilization. As Roosevelt saw it, continued progress meant the Plains Indians had to be removed in order to free up land for white settlers; the easiest way to do this was to destroy the buffalo herds upon which they depended. Neither the Indian nor the buffalo was well suited to survive the constant struggle for survival between species—or in the case of the Indian, between races and civilizations. The type of ecological warfare waged against the Plains Indians was quite familiar (and, it seems, unquestioningly accepted by Roosevelt), mirroring what had happened a century earlier to the Indians of the eastern woodlands, which Roosevelt described in *The Winning of the West*. Although he regretted the extinction of the buffalo, he also saw it as a harsh necessity of national progress:

While the slaughter of the buffalo has been in places needless and brutal, and while it is greatly to be regretted that the species is likely to become extinct . . . the extermination of the buffalo was the only way of solving the Indian question. As long as this large animal of the chase existed, the Indian simply could not be kept on reservations, and always had an ample supply of meat on hand to support them in the event of a war; and its disappearance was the only method of forcing them to at least partially abandon their savage mode of life. From the standpoint of humanity at large, the extermination of the buffalo has been a blessing. (1:191)

Without such a strong faith in the linear progress of history, founded on a belief that frontier life marked a "primitive stage of existence" that must "pass away before the onward march of our people" (292), it is unlikely that Roosevelt could have been so sanguine in his evaluation of what was happening to the west and to the frontier life he valued so highly.

Although Roosevelt's early books on ranch life and hunting reflect relatively little in the way of conservationist sentiment beyond the occasional note of regret about the diminishment of formerly plentiful numbers of game animals, his interest in hunting did lead in 1885 to the creation of the Boone and Crockett Club, one of the nation's first conservation organizations. Roosevelt and George Bird Grinnell, co-founders of the club, believed that it was incumbent upon sportsmen to work to protect game animals in particular and natural resources generally. During Roosevelt's tenure as president of the club (1888–1893), the two fought to preserve the game animals of the Yellowstone Park and published a number of books on nature and the outdoors. The club's membership rolls included such prominent names as Henry Cabot Lodge, Elihu Root, Francis Parkman, Owen Wister, and William T. Hornaday, which undoubtedly contributed to the image of elitism that plagued the group. Many commentators have noted the importance of sportsman's clubs in the early days of the conservation movement (although John Muir thought they were a "farce"); Stewart Udall, for one, refers to the Boone and Crockett Club's "outstanding contribution to our legacy of wild things" (149).[8]

Conservationist rhetoric appeared with increasing frequency in Roosevelt's later works, most notably in *Outdoor Pastimes of an American Hunter* (1905), which calls for better game laws, including bag limits; stricter enforcement; and the end of "unsportsmanlike" hunting practices. The "old days of wasteful plenty are gone forever" (52), Roosevelt declared, adding that any hunter not willing to observe reasonable restrictions on the amount of game taken "should be made to by the state" (89). As for himself, as "we grow older I think most of us become less keen about that part of the hunt which consists in the killing. I know that as far as I am concerned I have long gone past the stage when the chief end of a hunting trip was the bag" (58). Although hunting was too important a component of the strenuous life to be abandoned, Roosevelt was keenly aware how important it was for him—particularly since he was now president—to emphasize responsible game management practices.

Roosevelt's first real opportunity to incorporate conservation into public policy came in 1899, after his election as governor of New York State. As he noted in his autobiography, "All that later I strove for in the nation in connection with conservation was foreshadowed by what I strove to obtain for New York when I was governor; and I was already working in connection with Gifford Pinchot and Newell"[9] (284). Roosevelt's partnership with Pinchot resulted in some of the most important conservation achievements of the early twentieth century, but it also eventually splintered the conservation movement into two factions: the preservationists, with John Muir serving as the rhetorical and spiritual leader of that group; and the conservationists, or "wise use" advocates, led by Pinchot.[10] In contrast to George Perkins Marsh, who stressed utilitarian reasons for forest preservation out of a conviction that esthetic reasons were a given, Pinchot approached conservation from an economic viewpoint not simply because it was an effective rhetorical strategy but because he felt that other rationales were irrelevant. Forests were to be protected so that they would be available for use in perpetuity, providing "the greatest good to the greatest number of people."

In his first annual message as New York State's governor, in 1899, Roosevelt said little regarding the forests and wildlife of the state. Other than the state forest reserve program, which Roosevelt declared would stand as "a monument to the wisdom of its founders" (15:21–22), there is a surprising dearth of any substantive conservation programs. However, his message the following year enunciated a comprehensive policy toward the state's forests, asserting that their preservation "is of the utmost importance to the state" (15:53). Preserving the forests and the game animals they supported was important because "Hardy outdoor sports, like hunting, are in themselves of no small value to the national character, and should be encouraged in every way"; even more importantly, the preservation of the forests served a "more immediate and practical end." As George Perkins Marsh had done in *Man and Nature*, Roosevelt argued that the effect of forests on the watershed was an important enough reason to preserve them: "A primeval forest is a great sponge which absorbs and distills the rain-water; and when it is destroyed the result is apt to be an alternation of flood and drought" (54). The forest management philosophy of Gifford Pinchot becomes evident in the following passage:

We need to have our system of forestry gradually developed and conducted along scientific principles. When this has been done it will be possible to allow marketable lumber to be cut everywhere without damage to the forests—indeed, with positive advantage to them; but until lumbering is thus conducted, on strictly scientific principles no less than upon principles of the strictest honesty toward the State, we cannot afford to suffer it at all in the State forests. . . . Ultimately the administration of the State lands must be so centralized as to enable us definitely to place responsibility in respect to everything concerning them, and to demand the highest degree of trained intelligence in their use. (15:54)

The two prime tenets (the first formulated primarily by Roosevelt, the second by Pinchot) that would guide Roosevelt's conservation program during his presidency are already manifested here: conservation of forests and game animals in order to perpetuate the strenuous life, the foundation of national character; and the wise use and scientific management of natural resources.

In his public pronouncements and even in his outdoor books, Roosevelt spent little time dwelling on the spiritual or esthetic rationales for preserving wilderness, but concentrated almost exclusively on the pragmatic reasons. His first Annual Message as president, delivered in 1901, spoke of the growing public appreciation of "the value of forests," but there is no hint in his message that this value consisted of anything more than their role as potential resources to be exploited. Roosevelt extolled their part in "the creation and maintenance of national wealth," and justified his forest reserve policy by emphasizing economics:

> The fundamental idea of forestry is the perpetuation of forests by use. Forest protection is not an end of itself; it is a means to increase and sustain the resources of our country and the industries which depend upon them. The preservation of our forests is an imperative business necessity. (15:101–103)

Throughout his presidency Roosevelt emphasized that the nation's natural resources, including the forests, were to be managed and used: "It is the cardinal principle of the forest-reserve policy of this Administration that the reserves are for use. Whatever interferes with the use of their resources is to be avoided by every possible means" (15:235).

In his final Annual Message, in 1908, Roosevelt made his most extensive defense of the forest reserve policy, perhaps intending to impress the importance of this program upon his successors in office. In words that made it clear that the forests should be managed so as not to preclude future generations from their benefits he declared, "If there is any one duty which more than another we owe it to our children and our children's children to perform at once, it is to save the forests of this country, for they constitute the first and most important element in the conservation of the natural resources of this country" (15:517). He then described the damage that had already been done to the soil and waters of the United States as a consequence of "reckless deforestation," detailing the problems caused by deforestation in the Levant and in Asia, and drawing an explicit parallel to the situation in the United States:

> What has thus happened in northern China, what has happened in central Asia, in Palestine, in North Africa, in parts of the Mediterranean countries of Europe, will surely happen in our country if we do not exercise that wise forethought which should be one of the chief marks of any people calling itself civilized. Nothing should be permitted to stand in the way of the preservation of the forests, and it is criminal to permit individuals to purchase a little gain for themselves through the destruction of forests when this destruction is fatal to the well-being of the whole country in the future. (15:523)

Although no direct reference was made in this speech to George Perkins Marsh, it is interesting that Roosevelt, consciously or not, chose the same case histories of deforestation that Marsh had drawn on in *Man and Nature* (adding the timely example of China). Even more noteworthy is the fact that Roosevelt drew the same conclusions from these cases as had Marsh, and would no doubt have agreed with him that forest protection was one of "the most obvious of the duties which this age owes to those that are to come after it" (Marsh 279).

But while Roosevelt considered forest and water conservation to be "perhaps the most vital internal questions of the United States" (104), there were also intimations that he felt conservation should be something more than simply resource management or protection for the national psyche. His first Annual Message contained a proposal for certain forest reserves to be set aside as "preserves for the wild forest creatures," a concept that had more in common with Thoreau's idea of national forest preserves than with Pinchot's:

Some at least of the forest reserves should afford perpetual protection to the native fauna and flora, safe havens of refuge to our rapidly diminishing wild animals of the larger kinds, and free camping-grounds for the ever-increasing numbers of men and women who have learned to find rest, health, and recreation in the splendid forests and flower-clad meadows of our mountains. The forest reserves should be set apart forever for the use and benefit of our people as a whole and not sacrificed to the short-sighted greed of a few. (104)

Although the primary function of such reserves was to provide "rest, health, and recreation" for the American people, Roosevelt was willing to entertain the notion that they could serve a purpose less directly related to the national welfare.

Unfortunately for Roosevelt's future reputation as an environmentalist, his "resource management" arguments for conservation, while politically effective, tend to obscure the fact that there was an esthetic element to the strenuous life after all. As the descriptions of western life in his outdoor books indicate, Roosevelt was not blind to natural beauty, and in one of his later works, *A Book-Lovers Holidays in the Open* (1916), he made clear that the beauty of nature alone could be ample justification for preserving it:

A grove of giant redwoods or sequoias should be kept just as we keep a great and beautiful cathedral. The extermination of the passenger-pigeon meant that mankind was just so much poorer; exactly as in the case of the cathedral at Rheims. And to lose the chance to see frigate-birds soaring in circles above the storm, or a file of pelicans winging their way homeward across the crimson afterglow of the sunset, or a myriad terns flashing in the bright light of midday as they hover in a shifting maze above the beach—why, the loss is like the loss of a gallery of the masterpieces of the artists of old time. (3:377)

Although John Muir might have questioned Roosevelt's analogy of the extinction of the passenger pigeon with the destruction of a cathedral (after all, a cathedral, however beautiful, can be rebuilt; a species, once extinct, is gone forever), he would probably have been pleased to see that on at least some level his message about nature's spiritual and esthetic value had not been lost on Roosevelt. Roosevelt's love of birdwatching also contributed to the creation of the nation's first federal wildlife refuge. In 1903, a group of ornithologists approached him regarding possible federal protection for an endangered breeding ground on Pelican Island in Florida. Roosevelt asked his aides if there was any law that would prevent him from doing so, and when he was assured that there was not, he stated: "Very well, then I so declare it" (Cutright 178).

During the seven and one-half years of his presidency Roosevelt firmly established conservation as a national priority that all future presidents would have to address in one form or another. Realizing that it would be difficult to get many of his forest protection bills through Congress, he made liberal use of executive authority to set aside immense tracts of forested land—43 million acres in 1907 alone (20:395). He created several new national parks and signed the American Antiquities Act of 1906 into law, a measure that enabled him to protect eighteen sites, including the Grand Canyon, as national monuments. Roosevelt's work on behalf of forest protection continued unabated during his last year in office, and in May 1908 he convened a national conference on conservation at the White House. Although Gifford Pinchot managed to exclude many preservationists, including John Muir, from the conference, it was well received both by the participants and by the general public. Roosevelt convened a similar conference of North American countries in February 1909, shortly before he was to leave office, where the participants agreed on a declaration of principles including a resolution that "all nations should be invited to join together in conference on the subject of world resources, and their inventory, conservation, and wise utilization" (20:401–402). This plan for the first "earth summit" never came to fruition as Roosevelt left office the following month, entrusting his conservation program to his hand-picked successor, William Howard Taft.[11]

Given his record as president and his keen interest in natural history, Roosevelt's place as the nation's first "environmental president" should be secure; however, in a sense his modern reputation as an environmentalist is undermined by his own writings, with their emphasis on hunting and the strenuous life, which now seem like quaint echoes of the nineteenth century. Perhaps even more important was the role he played in factionalizing the conservation movement, a process that accelerated almost immediately after his term as president ended. While it may be ungenerous to blame Roosevelt for the split between Muir's preservationists and Pinchot's "wise use" partisans, he was probably the only one in a position to bridge the gap between the two groups and he failed to do so. Throughout Roosevelt's terms as governor and president, Pinchot's influence on conservation policy was so strong that competing voices would often find it difficult to get the president's ear. In a tribute to Muir written shortly after his death, Roosevelt himself noted this fact, gleefully recounting an incident in which Muir inadvertently gave him a letter wherein a correspondent of Muir's complained that Roosevelt was "altogether too much under the influence of that creature Pinchot" (11:290). This antagonism toward Pinchot by preservationists was quite common; J. Horace McFarland, the head of the American Civic Association and a devoted protector of Niagara Falls, commented that Pinchot "does not see beyond the end of the lumber pile in regard to the forests" (Fox 132), an assessment Muir probably would have seconded.

Although Roosevelt gave Pinchot much of the credit for the conservation achievements of his administration, the question of exactly where Roosevelt himself stood on the issue of wise use vs. preservation is more ambiguous than it would at first appear. Pinchot, for one, was worried that Muir's lobbying would tilt the president toward the preservationist side, and so he did not invite Muir to the national conference on conservation in 1908. In 1903 Roosevelt toured the Yellowstone Park with Burroughs and the Yosemite with Muir. He later said of this trip in his autobiography, "I shall always be glad that I was in the Yosemite with John Muir and in the Yellowstone with John Burroughs" (20:312). In describing the trip in *Outdoor Pastimes of an American Hunter*, he wrote that such places should be protected "not merely for the sake of preserving the forests and the water, but for the sake of preserving all its beauties and wonders unspoiled by greedy and short-sighted vandalism" (3:89–90). This esthetic appreciation of nature is largely absent from his earlier works and may have come about under the influence of Muir, and of Burroughs as well, to whom *Outdoor Pastimes* was dedicated.[12]

However, even this tentative movement toward an esthetic appreciation of nature is tinged with political overtones, as was his occasionally ostentatious friendship with Burroughs. Roosevelt had been under increasing fire in the press for his hunting, and the impression of him as a "game-butcher" was growing. His trip to Yellowstone Park was originally supposed to include hunting, but widespread criticism in the newspapers stymied this plan, and Roosevelt contented himself with a more political bag—being seen in the company of the nation's most beloved naturalists. Burroughs had already done the president a considerable political service in the form of an essay in the May 1903 edition of the *Atlantic Monthly* entitled "Real and Sham Natural History," in which he attacked what he believed was inaccurate and overly sentimental work by a number of nature writers. "Sentimental" nature writing presented a real political danger to Roosevelt—if the American public began to believe animals possessed many of the same traits as humans (such as the ability to feel pain and fear), his hunting would seem even more reprehensible to many people. Roosevelt's appreciation for Burroughs's stance was expressed in his dedication to *Outdoor Pastimes of an American Hunter*, where he cites with approval Burroughs's "warfare against the sham nature-writers."

For over two years the "nature-faker" controversy raged in various periodicals while Roosevelt maintained a discreet presidential silence, content to let Burroughs and his philosophical allies carry on the battle with the offending writers. Finally, he could maintain his silence no longer and weighed in with an interview in the June 1905 issue of *Everybody's Magazine* in which he vigorously attacked William Long and Jack London, among others, for their skewed pictures of natural history, saying, "I don't believe for a minute that some of these nature writers know the heart of a wild thing." Long's response was to compose an open letter to the president that ran in a number of newspapers. In the letter he deftly turned Roosevelt's own words against him:

I find after carefully reading two of his big books that every time Mr. Roosevelt gets near the heart of a wild thing he invariably puts a bullet through it. From his own records I have reckoned a full thousand hearts which he has thus known intimately. In one chapter alone I find that he violently gained knowledge of 11 noble elk hearts in a few days. (Cutright, *TRN* 135)

While Long and the nature-faking issue have largely receded into the background, the image of Roosevelt as a gun-crazy hunter is one that has lingered.[13]

While Roosevelt's political accomplishments in furthering conservation make him one of the most important figures in the history of environmental reform in America, his literary and philosophical legacy in this area, as we have seen, is obviously on shakier ground. His rhetoric, unlike Muir's or Thoreau's, seldom transcends specific issues to speak to the heart of the modern environmentalist. Thus his work often seems dated—the strenuous life, so central to his books and his philosophy, is now far more likely to take place in a health club than in the mountains on a hunting expedition. Still, Roosevelt's literary contribution to the early conservation movement cannot be wholly discounted since, like Burroughs, his books helped attract a constituency for conservation from among a class of people, such as sportsmen, who would have remained largely unmoved by Muir's rhetoric. Even for today's readers, Roosevelt's stories of hunting and outdoor life retain a relevance, as Aldo Leopold detailed in *A Sand County Almanac*, noting that his writings demonstrate the value of "a code of sportsmanship, a self-imposed limitation on sport . . . a distinctively American tradition of self-reliance, hardihood, woodcraft, and marksmanship" (213). His literary contribution to the conservation movement was expressing "this intangible American tradition in words any schoolboy could understand," thereby helping to create "cultural value by being aware of it, and by creating a pattern for its growth" (214). It may be a testament to Roosevelt's success in fostering the growth of the now generally accepted cultural value of environmentalism that his own notions of conservation seem like a vestige of the past.

Notes

1. In *Deep Ecology: Living as if Nature Mattered*, Devall and Sessions trace the movement to ecological consciousness in America through political reform. They note that "until recently, modern people who considered themselves enlightened on the human/Nature relationship have thought of themselves as *conservationists*" (56). While many present day conservationists still consider themselves to be "enlightened" on the issue of the human/nature relationship, mainstream environmentalists now generally consider resource management, the centerpiece of conservationism, to be merely one factor (and many would say a minor factor) in the issue of wilderness protection.

2. Most of the "Roosevelt Museum" was donated to the Smithsonian Institution in 1882 (Cutright, *TRN* 34).

3. Aloysius Norton asserts in his critical biography of Roosevelt that his descriptions of characters he met while in the west rival the fictional creations of Jack London (98).

4. "The buffalo are now gone forever, and the elk are rapidly sharing their fate; but antelope and deer are still quite plenty, and will remain so for some years; and these are the common game of the plainsman" (1:24).

5. In planning for his trip to Yellowstone Park in 1903, Roosevelt wrote to the park superintendent to inquire about the possibility of hunting mountain lions there, "on the supposition that they are 'varmints' and are not protected" (Cutright, *TRN* 105).

6. "[T]he incoming of the settlers and the driving out of the Indians have left the ground clear for the trappers to work over [the beaver meadows] unintermittently, and the extinction of the beaver throughout the plains country is a question of but a short time. . . . We have trapped (or occasionally shot) on the ranch during the past three years several score of beaver; the fur is paler and less valuable than in the forest animal" (Roosevelt 1:35).

7. Parkman's influence on Roosevelt in this instance, as in many others, was considerable—and the parallels between the two are intriguing. They both had an insatiable interest in the wilderness and in American history, and despite childhood infirmities they engaged in vigorous outdoor activities such as lengthy camping trips in the woods of the northeastern United States. Both graduated from Harvard (which isn't so unusual considering the fact that they both came from well-established eastern families of some means), and wrote histories of North America in which the influence of the wilderness was highlighted. The Byronic epigraph from *The Oregon Trail* could well serve as the epigraph for any of Roosevelt's books on outdoor life: "Let him who crawls enamor'd of decay, / Cling to his couch, and sicken years away; / Heave his thick breath, and shake his palsied head; / Ours— the fresh turf, and not the feverish bed."

8. For a well-researched if partisan history of the impact of hunting organizations on conservation policy, see John F. Reiger, *American Sportsmen and the Origins of Conservation*.

9. Frederick H. Newell, an aide to Pinchot and Roosevelt, would later found the U.S. Reclamation Service.

10. Born to a wealthy Connecticut family in 1865, Gifford Pinchot studied forestry in Europe and served as chief of the United States Forest Service from 1898 to 1910. Following his dismissal from this position by William Howard Taft (an act that was one of the primary reasons for Roosevelt's falling out with his successor), he went on to teach forestry at Yale University and serve as governor of Pennsylvania. He wrote a number of works on forestry, including *A Primer of Forestry* (1899), *The Fight for Conservation* (1909), and *The Training of a Forester* (1914).

11. For a description of Roosevelt's conservation conferences see Paul Cutright, *Theodore Roosevelt the Naturalist*, 179–83.

12. The possibility exists that in some ways Roosevelt may have suffered from a bit of esthetic myopia—as Paul Cutright notes, despite his acute powers of observation, Roosevelt never commented upon "the beauty or plumage of a bird, not even of an oriole, tanager, or grosbeak" (*TMN* 81). Oddly enough, during his tour of the Yosemite with Muir, Roosevelt noted a similar blindspot in his host, commenting upon Muir's curious (at least to Roosevelt) lack of interest in bird songs (11:289). Cutright suggests that Muir's lack of interest in birds may have been one of the reasons why Roosevelt, while he liked and admired Muir, never quite warmed to him the way he did to Burroughs (*TRN* 117). Of course, this may just as well be attributed to the fact that Muir and Roosevelt were both inveterate talkers, while Burroughs was more of a listener.

13. Ironically, Roosevelt is often guilty of ascribing human motivations or traits to animals, thereby committing the error of "nature-faking" himself. For instance, he claims that elk will sometimes "destroy [porcupines] in sheer wantonness" (2:138), and at one point characterizes a rutting moose as "evidently in a most evil rage and bent on man-killing" (3:398). This tendency occasionally results in some florid prose, as when he describes a treed cougar as "the big horse-killing cat, the destroyer of the deer, the lord of stealthy murder, facing his doom with a heart both craven and cruel" (3:199).

Works Cited

Allin, Craig W. *The Politics of Wilderness Preservation.* Westport, CT: Greenwood Press, 1982.
Cutright, Paul. *Theodore Roosevelt: The Making of a Naturalist* (cited as *TMN*). Urbana and Chicago: University of Illinois Press, 1985.
——. *Theodore Roosevelt the Naturalist* (cited as *TRN*). New York: Harper & Brothers, 1956.
Devall, Bill, and George Sessions. *Deep Ecology: Living as if Nature Mattered.* Salt Lake City: Peregrine Smith Books, 1985.
Gould, Lewis L. *The Presidency of Theodore Roosevelt.* Lawrence: University Press of Kansas, 1991.
Rev. of *Hunting Trips of a Ranchman* by Theodore Roosevelt. *New York Times,* July 13, 1885.
Leopold, Aldo. *A Sand County Almanac.* 1949. New York: Ballantine Books, 1966.
Marsh, George Perkins. *Man and Nature.* 1864. Cambridge: The Belknap Press of Harvard University Press, 1965.
Mitchell, Lee Clark. *Witnesses to a Vanishing America: The Nineteenth-Century Response.* Princeton: Princeton University Press, 1981.
Morris, Edmund. *The Rise of Theodore Roosevelt.* New York: Coward, McCann & Geoghegan, Inc., 1979.
Nash, Roderick. *Wilderness and the American Mind.* 1967. New Haven: Yale University Press, 1982.
Norton, Aloysius. *Theodore Roosevelt.* Boston: Twayne Publishers, 1980.
Oelschlaeger, Max. *The Idea of Wilderness.* New Haven: Yale University Press, 1991.
Parkman, Francis. *The Oregon Trail.* 1849. New York: Penguin Books, 1982.
Reiger, John F. *American Sportsmen and the Origins of Conservation.* New York: Winchester Press, 1975.
Roosevelt, Theodore. *The Works of Theodore Roosevelt: National Edition.* 20 vols. New York: Charles Scribner's Sons, 1926. Vol. 1, "Hunting Trips of a Ranchman" and "Ranch Life and the Hunting Trail" (1885, 1888). Vol. 2, "The Wilderness Hunter" and "Outdoor Pastimes of an American Hunter I" (1893). Vol. 3, "Outdoor Pastimes of an American Hunter II" and "A Book-Lover's Holidays in the Open." Vol. 4, *African Game Trails* (1909). Vols. 8, 9, *The Winning of the West* (1889). Vol. 11, "The Rough Riders" and "Men of Action." Vol. 12, *Literary Essays.* Vol. 13, "American Ideals" and "The Strenuous Life" (1897, 1899). Vol. 15, *State Papers as Governor and President 1899–1909.* Vol. 20, *Theodore Roosevelt: An Autobiography* (1913).
Savage, Henry, Jr. *Lost Heritage.* New York: William Morrow and Company, 1970.
Udall, Stewart L. *The Quiet Crisis and the Next Generation.* 1963. Salt Lake City: Peregrine Smith Books, 1988.
Worster, Donald. *Nature's Economy: A History of Ecological Ideas.* New York: Cambridge University Press, 1977.

CHAPTER 7

Alone in a World of Wounds: The Question of Audience in *A Sand County Almanac*

> *One of the penalties of an ecological education is that one lives alone in a world of wounds. Much of the damage inflicted on land is quite invisible to laymen. An ecologist must either harden his shell and make believe that the consequences of science are none of his business, or he must be the doctor who sees the marks of death in a community that believes itself well and does not want to be told otherwise.*
> —Aldo Leopold, "The Round River—A Parable," 1953

Since his death in 1949—and especially since the explosive growth of the environmental movement in the 1960s and 1970s—Aldo Leopold has been elevated to the pantheon of eco-prophets, sharing that honor with such figures as Henry David Thoreau, John Muir, and Rachel Carson. There is ample reason for praise. Leopold distinguished himself in a series of conservation careers: as a forester, game management expert, activist and co-founder of the Wilderness Society, teacher, and writer. His work in any one of these areas is enough to warrant notice, but the primary reason for this enshrinement is a slim book of essays published shortly after his death. *A Sand County Almanac*[3] has many of the same talismanic qualities as *Walden*, and its posthumous publication cemented Leopold's reputation as an ecological visionary. Wallace Stegner has called *A Sand County Almanac* "one of the prophetic books, the utterance of an American Isaiah" (233), and diverse commentators ranging from Rene Dubos to Dave Foreman have agreed with this assessment.[2]

There are numerous reasons for the popularity of *A Sand County Almanac* and for the profound effect that it has had on millions of readers, particularly those with an interest in protecting the environment. Leopold's anecdotes and observations from a lifetime spent in the field are vividly written, conveying a singular clarity of vision concerning mankind's relationship with the land. His writing style is concise and "literary," yet largely devoid of literary pretension. The tone throughout the work is that of a credible, thoughtful, and sympathetic guide intent upon explaining, as he writes in the foreword, why "the company is out of step" with the natural world. Finally, and perhaps most significantly, there is "The Land Ethic," a blueprint for human-land relations that provides the most succinct standard yet formulated for a biotic approach to land use: "A thing is right when it tends to preserve the integrity, stability, and beauty of the biotic community. It is wrong when it tends otherwise" (240). Like Einstein's theory of relativity, the elegant simplicity of Leopold's formula for healthy man-land relations partially masks a sea change in perception.[3] The nuances of the land ethic have been vigorously debated, but at the very least, two things are certain: first, at the time Leopold formulated his land ethic the idea of a biotic, or nonanthropocentric, community was a radical one in the sense that only a tiny minority was aware of the concept and even fewer subscribed to it; and second, *A Sand County Almanac* stands alone as a succinct, accessible, and persuasive argument in favor of a biotic/earth-centered philosophy of land use.

Still, despite the undeniable fact that *A Sand County Almanac* has reached and inspired untold numbers of environmentalists, these were not the intended audience—at least not the primary audience—that Leopold envisioned when he wrote the book. His actual audience, as he saw it, was sportsmen, landusers such as farmers and ranchers, and the professional class of conservationists of which he was himself a member. Many critics have failed to note this threshold issue, with the result that a number of essays suffer from a fundamental misapprehension of the purpose of the book. For instance, as J. Baird Callicott points out, some of the critics who have approached *A Sand County Almanac* from the standpoint of ethics and philosophy have castigated Leopold for his "extremely condensed prose style in which an entire conceptual complex may be conveyed in a few sentences . . . [and]

his departure from the assumptions and paradigms of contemporary philosophical ethics" (*IDLE* 76). Others have criticized his support of hunting, with one animal rights advocate calling Leopold's emphasis on the biotic community rather than individuals "environmental facism" (Regan 361–62). Such criticism fails to take into account that Leopold's audience was one that would recognize and appreciate his down-to-earth rhetoric far more than it would the "assumptions and paradigms" of academic philosophers, and one that, for the most part, would dismiss the position of animal rights activists on hunting as eccentric extremism. The land ethic and the other elements of Leopold's ecological ethics were the long considered principles of an expert in the field; nevertheless it was vitally important to Leopold that they be communicated to those who, while perhaps not experts or committed conservationists, were certainly "in the field." In short, *A Sand County Almanac* was primarily addressed to private landusers whose actions affected both the human and the biotic community at large. As Leopold wrote in "The Ecological Conscience": "Despite nearly a century of propaganda, conservation still proceeds at a snail's pace; progress still consists largely of letterhead pieties and convention oratory. On the back forty we still slip two steps backward for each forward stride" (*SCA* 222).

In addition to farmers and other landowners, Leopold believed it was vitally important to the success of environmental reform to enlist the support of the "shotgun vote," the sportsmen whose growing influence he had first noted in "Ten New Developments in Game Management" (1925). Leopold was himself an avid hunter; his first exposure to conservation ethics came through his father, who adhered to a strict personal code of sportsmanship. Despite the fact that such legislative measures as restricted seasons, bag limits, and similar game laws had not yet been passed into law in the Leopolds' home state of Iowa, Carl Leopold had noted the decreasing numbers of migratory waterfowl and had already restricted his own hunting accordingly (Meine 18). Carl Leopold was a great admirer of Theodore Roosevelt, whose books on hunting and outdoor life also appealed to Aldo. Like Roosevelt, Aldo Leopold subscribed to Frederick Jackson Turner's thesis on the cultural value of the frontier and strongly believed that outdoor activities such as hunting and camping were an important means of building character both on the individual and on the national level. Hunting in particular, argued Leopold, had the "peculiar virtue" of a set of ethics that for the most part were dictated from within: "the hunter ordinarily has no gallery to applaud or disapprove of his conduct. Whatever his acts, they are dictated by his own conscience, rather than by a mob of onlookers. It is difficult to exaggerate the importance of this fact" (*SCA* 197).

In "Game and Wild Life Conservation" (1932), Leopold sought to bridge the gap between protectionists and sportsmen, pointing out that there were many "intergrades like myself, who share the aspirations of both" (*RMG* 164), and arguing that they had a great deal of common ground in causes such as game protection. As Curt Meine has noted, Leopold had a "penchant for acting as peacemaker in conservation quarrels . . . between aesthetes and utilitarians, sportsmen and preservationists, academics and outdoorsmen, managers and observers," in part because he was, in a way, all of these things himself (280). Leopold's sense of himself as a bridge between differing factions of the conservation movement is an important part of the background of *A Sand County Almanac* and of many of the other essays he wrote throughout his career.

In 1906, Leopold's interest in the outdoors led him to enroll in the Yale School of Forestry, established by the Pinchot family just six years earlier. While at Yale Leopold was indoctrinated in the "wise use" principles of forestry propounded by Roosevelt's chief forester, Gifford Pinchot. A career limited to calculating board feet of lumber and the grazing capacity of rangeland, however, held limited appeal for Leopold and, while still at forestry school, he had already declared an aversion to being a "tie-pickler or a timber tester" (Meine 234). As early as the 1920s, in his arguments in favor of establishing the Gila Wilderness Area in New Mexico, Leopold was stretching Pinchot's economic rationale for wilderness protection as far as it would go, and there is little indication in his later writings that he believed economic factors to have more than a superficial validity in land use decisions. Even after he had ceased to believe that the profit motive was an important part of ecological decision making, however, Leopold would refer to it whenever it was rhetorically convenient to do so. As he wrote in "Game and Wildlife Conservation" (1932), "if we want Mr. Babbitt to rebuild outdoor America, we must let him use the same tools wherewith he destroyed it. He knows no other" (*RMG* 166).

Following his graduation from forestry school in 1909, Leopold joined the U.S. Forest Service and was assigned to the Apache National Forest in Arizona, the first of a series of assignments in the southwest over the next fifteen years. It was while stationed in the southwest that he first became aware of how important it was to gather the support of landusers if meaningful conservation reform were to be successfully implemented. As a professional forester, the bulk of Leopold's duties revolved around extracting resources from the national forests, but it wasn't long before his interest in hunting led to what would become the main focus of his work over the next twenty-five years—game conservation. Although Leopold gave game protection a higher priority than did some of his superiors in the Forest Service, it appears that two bureaucratic staples—interagency competition and the desire for reflected glory—worked to provide him the latitude he needed. In December 1915 Leopold founded *The Pine Cone*, the official bulletin of the Albuquerque (later the New Mexico) Game Protective Association, and an invaluable forum for many of his earliest essays on game protection and related issues.[4] Many of the articles Leopold wrote for *The Pine Cone* and for specialized publications such as *The Journal of Forestry* and *The Condor*, an ornithological journal, were drily technical, as indicated by titles such as "Determining the Kill Factor for Blacktail Deer in the Southwest," "Weights and Plumage of Ducks in the Rio Grande Valley," and "Grass, Timber, and Fire in Southern Arizona"; but the process of researching and writing these technical essays undoubtedly contributed to the careful attention to detail that characterizes Leopold's work.

Many of the early essays were attempts to convince groups such as ranchers, sportsmen, and fellow foresters of the value of habitat protection for game animals, emphasizing issues such as fire and erosion control, and wilderness protection, as well as other less salutary measures such as the eradication of predators. Leopold was also in great demand as a lecturer, and this experience was nearly as important to the development of his rhetorical skills as was his writing. Speaking in front of audiences who could not be counted on to be particularly receptive to his ideas, Leopold learned how to preach an unfamiliar, even seemingly radical, ecological gospel to audiences other than those in the conservation choir. Like the experienced speaker who starts off with a joke or humorous anecdote in order to break the ice with his audience, Leopold realized how important it was to "shoot the breeze for a while before coming down to business," as Curt Meine has pointed out (262). In some of Leopold's earlier essays, he tended to cross the line between political exhortation and ideological shrillness; his experience in addressing a live audience that would respond instantly and negatively went a long way toward ameliorating this.

Over the next quarter of a century, Leopold's focus slowly changed as game management became wildlife protection, and finally ecology. Just as Leopold's ideas on ecology and habitat protection were evolving during this period, so were his notions of how best to effect meaningful change in the way Americans approached land relations. In early essays, Leopold usually reflected the Progressive notion that enlightened government policies guided by sound scientific principles were the key to reform, but as time went by, an awareness of the limits of governmental power to effect change emerged in his writing. Whether it was his experience chasing poachers around the forests of the southwest or his sometimes frustrating efforts to enlist the support of cantankerous ranchers, sportsmen, and politicians in the cause of conservation, or simply an innate Jeffersonian skepticism of government, Leopold's faith in legislation as the main answer to conservation problems had already begun to diminish by the 1920s. By the time he wrote his pivotal essay (and precursor to "The Land Ethic"), "The Conservation Ethic," in 1933, Leopold believed that legislation was but one dimension of a threefold process that included self-interest and ethics. As Leopold caustically remarked in "Land Use and Democracy" (1942), to deal only with "bureaus, policies, laws, and programs" reminded him of how his dog behaved when confronted with a bigger dog: "Instead of dealing with the dog, he deals with a tree bearing his trademark. Thus he assuages his ego without exposing himself to danger" (RMG 295).

Such abuses as the destructive farming practices that had substantially contributed to the great dust storms of the 1930s convinced Leopold that government action alone—even the avalanche of government programs that accompanied the New Deal—was insufficient to enact the necessary change in land use practices. Instead, in a conservation equivalent of a "hearts and minds" policy, Leopold argued that education in ecological ethics was the key to persuading landusers to assume real responsibility in conservation policy on an individual and community level. Leopold had few illusions about what it would take to instill such a sense of responsibility in the farmers, the hunters, and the fishermen, who had a significant impact on the land and its creatures. As he wrote in "The Ecological Conscience": "It has required 19 centuries to define decent man-to-man conduct and the process is only half done; it may take as long to evolve a code of decency for man-to-land conduct" (RMG 345). As Leopold saw it, the problem was how to get the necessary process of defining such a standard of man-to-land relations started. On a personal level, one of the ways he did this was to buy a "sand farm in Wisconsin, first worn out and then abandoned by our bigger-and-better society" (SCA ix–x). The visits of Leopold and his family to "the shack" and their efforts to reclaim the farm's exhausted land enabled him to put his ecological principles into practice, and would later go to form the narrative heart of *A Sand County Almanac*.

The challenge Leopold faced in putting together the work that became *A Sand County Almanac* was how to present a lifetime of ecological study, observation, and conclusions in a volume that—unlike his earlier, academic text *Game Management* (1933)—would speak to sportsmen and farmers as well as conservation professionals. In December 1941, Alfred A. Knopf Publishers approached Leopold with a proposal for "a personal book recounting adventures in the field . . . warmly, evocatively, and vividly written . . . a book for the layman . . . [with] room for the author's opinions on ecology and conservation . . . worked into a framework of actual field experience" (cited in Ribbens 93). Over the next several years, Leopold and Knopf sought unsuccessfully to strike a mutually agreeable balance between Knopf's desire for a "personal book recounting adventures in the field" and Leopold's "opinions on ecology and conservation";[5] however, the revisions Leopold made in order to satisfy Knopf did result in a stronger narrative line and a more pronounced unity among the essays. The original plan called for Albert Hochbaum, a former student of Leopold's to whom Leopold referred as "one of my intellectual anchors" (Meine 508), to provide sketches for the book. Despite the fact that Hochbaum's own work ultimately prevented him from this artistic contribution, his continuing editorial input significantly improved the overall rhetorical effectiveness of the work.

Most significantly, Hochbaum identified a problem with the tone of the work that may otherwise have prevented Leopold from reaching the very audience he sought most to influence. In a series of letters that passed between the two men, Hochbaum pointed out—and Leopold finally agreed—that there was a problem of "attitude" in the essays. In a letter of February 4, 1944, Hochbaum wrote:

There is one false note—the reader cannot help but gather that you believe your reaction is always the proper one and that it has been always so. Don't get me wrong; the lesson you wish to put across is the lesson that must be taught—preservation of the natural. Yet it is not easily taught if you put yourself above other men. (cited in Meine 454)

Hochbaum suggested that Leopold demonstrate that he had not always "smoked the same tobacco," by writing about an instance where he himself had been woefully mistaken in his earlier years, his role in the eradication of wolves in the southwest. The recommendation led directly to one of Leopold's finest essays, "Thinking Like a Mountain." In urging Leopold to detail his own past ecological "sins" in order to provide a contrast with his present beliefs, Hochbaum was suggesting a rhetorical strategem at least as old as Augustine's *Confessions*, and one that was particularly well suited to an audience at once conservative and fiercely independent. Instead of berating his readers for their ecological sins, Leopold used himself as an example, describing some of his own past sins and detailing how he came to see the error of his ways. As John Tallmadge has pointed out, *A Sand County Almanac*—in particular "Thinking Like a Mountain"—contains many of the elements of a traditional conversion narrative, which, "[l]ike all conversion stories . . . inspires emulation" (114). For Leopold, to merely convince farmers, sportsmen, and other landusers of the truth of his message was not enough—he had to convert them to his way of thinking and inspire them to action, on however small a scale.

"Thinking Like a Mountain" is by no means the only instance in *A Sand County Almanac* where Leopold alluded to Christian tradition in order to convey his message about the moral issues involved in land use decisions. Right from the start, the foreword refers to Part Three as "some of the ideas whereby we dissenters rationalize our dissent," a phrase subtly evoking a history of conscientious dissent that harkens back to Martin Luther, if not to Christ himself. The quasi-religious underpinning of *A Sand County Almanac* is further reinforced by what Leopold says in the next paragraph:

Conservation is getting nowhere because it is incompatible with our Abrahamic concept of land. We abuse land because we regard it as a commodity belonging to us. When we see land as a community to which we belong, we may begin to use it with love and respect. There is no other way for land to survive the impact of mechanized man, nor for us to reap from it the esthetic harvest it is capable, under science, of contributing to culture. (x)

Here Leopold explicitly extends the Christian concept of a "brotherhood of man" to that of a community comprised of the land and all of its inhabitants, a concept that is at the heart of the land ethic. Like John Muir, Leopold was adept at presenting new or unfamiliar concepts wrapped in comforting and familiar religious imagery. Also like Muir, Leopold's religious beliefs were probably akin to pantheism;[6] but, although not a member of any Christian sect, he was well versed in the Bible—see, for example, his 1920 essay "The Forestry of the Prophets"—and he skillfully drew on this source in his writing.

A lifetime of work in conservation had provided Leopold with an enviable set of credentials as an expert in the field, on which he drew heavily in making the case for his land ethic; but he also knew enough about his audience to take pains to avoid sounding like an expert come to tell other people how to live. His dealings with the ranchers and sportsmen of the southwest had undoubtedly given him first-hand experience of the anti-expert bias, and, in the unlikely event he had forgotten that lesson, the abuse he took in a campaign to persuade the

Wisconsin public that the state's deer population in the early 1940s had exceeded the land's capacity to support it would have reminded him.[7] The prose of *A Sand County Almanac*—even in "The Upshot," where he is dealing with some fairly abstract concepts—is intimate and straightforward, the honest observations of an expert who doesn't direct his comments over the head of his audience. Academics might speak in terms of biota and ethical paradigms, but Leopold knew that the majority of his readers would be far more interested in forests, fields, and streams, and that he had to strike a balance between the two approaches if he was to keep his audience interested in the former. Realizing that his concepts of land health as outlined in "The Upshot," including the land ethic itself, represented the most challenging aspect of the work, Leopold beckons the reader on with a self-deprecating, and perhaps somewhat disingenuous, disclaimer that only a "very sympathetic" reader will "wish to wrestle with the philosophical questions" of that section. The disclaimer is part of Leopold's carefully calculated lead-in to the system of ecological ethics that he sketches out in "The Upshot": after all, what reader would want to see himself as either unsympathetic or intellectually unsuited to wrestle with these philosophical questions?

As in his early public lectures, Leopold realized that here, too, it was necessary to "shoot the breeze" for a while before coming down to the real issues involved, and this is reflected in the structure of *A Sand County Almanac*. Parts One and Two are comprised primarily of anecdotes from the field, although, as John Tallmadge has pointed out (123), many of these stories are in fact parables that convey a message pertinent to the more explicit statements on ecology that come primarily in "The Upshot." For example, when Leopold gently mocks the mouse, "who knows that grass grows in order that mice may store it as underground haystacks," or the hawk, who "has no opinion why grass grows, but he is well aware that snow melts in order that hawks may again catch mice" (4), he is calling the human perspective into question as well. And when, a few pages later, he discusses a lightning storm that struck an oak near the shack, the lesson behind the previous series of sketches becomes clearer: "we realized that the bolt must have hit near by, but, since it had not hit us, we all went back to sleep. Man brings all things to the test of himself, and this is notably true of lightning" (8). The same proposition is discussed in "The Upshot," but this time it is stated more starkly: "Abraham knew exactly what the land was for: it was to drip milk and honey into Abraham's mouth. At the present moment, the assurance with which we regard this assumption is inverse to the degree of our education" (220).

Leopold's belief that ecological education and individual responsibility were the key to land health is at the heart of *A Sand County Almanac*, and of his later writings as well; and this focus again suggests that the book is directed at landusers as well as conservation professionals. Leopold had come to doubt his earlier and oft-stated belief that "Mr. Babbitt" could be induced to rebuild outdoor America by an appeal to his economic self-interest. As he wrote in "The Round River—A Parable," he hesitated to reject the profit motive as a means of restoring land to health, given its "prodigious achievements" in wrecking the land in the first place, but he now believed that the power of economic motivation had been overstated. The "ethical and esthetic premises which underlie the economic system" had their corollaries in ecological decisions, but these had been ignored thus far: "I think we have here the root of the problem. What conservation education must build is an ethical underpinning for land economics and a universal curiosity to understand the land mechanism. Conservation may then follow" (SCA 187).

In essay after essay, Leopold drives home the point that the individual landowner and landuser must be convinced that there are ethical dimensions to land use every bit as real as the ethics governing human relations. In "The Land Ethic," Leopold attacked as outmoded and, in the final analysis, unethical the old Pinchotian notion that economics should govern all land use:

When the private landowner is asked to perform some unprofitable act for the good of the community, he today assents only with outstretched palm. If the act costs him cash this is fair and proper, but when it costs only forethought, open-mindedness or time, the issue is at least debatable. . . . [A] system of conservation based solely on economic self-interest is hopelessly lopsided. It tends to ignore, and thus eventually to eliminate, many elements in the land community that lack commercial value, but that are (as far as we know) essential to its healthy functioning. It assumes, falsely, I think, that the economic parts of the biotic clock will function without the uneconomic parts. It tends to relegate to government many functions eventually too large, too complex, or too widely dispersed to be performed by government.

An ethical obligation on the part of the private owner is the only visible remedy for these situations. (230)

The sense of ethical obligation—"social stigma in the possession of a gullied farm, a wrecked forest, or a polluted stream" (158); as well as the sense of personal fulfillment that arose, as Leopold knew from his personal experience at the shack, from preserving or restoring land health, was far more important than the "letterhead pieties and convention oratory" to which the conservation movement had largely succumbed: "We tilt windmills in behalf of conservation in convention halls and editorial offices, but on the back forty we disclaim even owing a lance" (158).

The formation of ethical principles for use in ecological decision making is, Leopold says in "The Land Ethic," a challenge that has been "*so far* studied only by philosophers" (my emphasis), but it is clear that Leopold's purpose in *A Sand County Almanac* was to broaden the discussion of ecological ethics to include the general public. The book contains a number of anecdotes meant to inspire individual landowners to action, such as the efforts of two Wisconsin farmers to plant a tamarack grove on their land because they wished to reintroduce the "nearly extinct wildflowers of the aboriginal Wisconsin bogs" (188), or the cases of inspired amateurs whose nature studies both provided personal pleasure and made significant contributions to science (204). These examples are pointedly contrasted with the "worried farmer, his fertilizer bill projecting from his shirt pocket," who doesn't realize that the native prairie grasses he has tried so hard to eradicate were what made his land fertile in the first place (118), or the sportsmen and stockmen who wrangle over land rights on the winter range while the invasion of cheat grass diminishes the value of the land as graze for either cattle or game (157). The sheer number of such examples in the book makes clear that Leopold would have considered a discourse on the land ethic that was limited to philosophers and academics a profound failure, a wasted opportunity similar to countless others in conservation history.

Once we realize who comprised the intended audience for *A Sand County Almanac*, some of the seeming anomalies of the work (and others such as *Round River*) fall into place. For instance, as astute a commentator as Rachel Carson criticized Leopold for what she saw as his overemphasis on hunting (Meine 525); and indeed, some of his comments on the cultural value of hunting, particularly in the essays from *Round River* that were included in later editions of *A Sand County Almanac*, are overstated, to say the least.[8] Others, such as "Red Legs Kicking," where Leopold describes his "unspeakable delight when my first duck hit the snowy ice with a thud and lay there, belly up, red legs kicking" (*SCA* 121) seem almost calculated to appall the nonhunter. However, when one takes into account the fact that Leopold was seeking to move beyond "letterhead pieties," proclamations, and government agencies to a revolutionary change of ethical standards where it counted most, in the fields and forests outside the public domain, then such passages begin to fit. Furthermore, argued Leopold, there was room for all different kinds of nature lovers in the conservation movement, including sportsmen; in a sense, every conservationist was a hunter anyway, "because the wild things he hunts for have eluded his grasp, and he hopes by some necromancy of laws, appropriations, regional plans, reorganization of departments, or other form of mass-wishing to make them stay put" (258). *A Sand County Almanac* has been described as a "subversive" work,[9] and in Leopold's efforts to convince farmers, sportsmen, and other landusers that conservation was more than simply good business, he was in fact a classic subversive, quietly and effectively undermining the prevailing system from within, rather than ineffectually throwing rhetorical bombs from without.

Notes

1. Except as noted, when citing *A Sand County Almanac* I will be referring to the expanded edition published by the Oxford University Press in 1966, rather than the 1949 edition. The later compilation includes all of the essays included in the first edition as well as eight, including "The Round River— A Parable," that were originally published in 1953 under the title *Round River: From the Journals of Aldo Leopold*.

2. In *The God Within*, Dubos refers to *A Sand County Almanac* as "the Holy Writ of American conservationists" (156), and in a 1986 interview Foreman, co-founder of Earth First!, stated that it was "not only the most important conservation book ever written, it is the most important book ever written" (cited in Nash 231). Numerous other critics, including Stewart Udall, Holmes Rolston III, Joseph Des Jardins, Craig Allin, and others have also cited the importance of *A Sand County Almanac* in forming American environmental thought in the twentieth century.

3. For the most part, Leopold's ideas on ecology weren't particularly new, although his success in presenting them as a viable alternative to economically based land use was. The roots of his land ethic can be traced back through a line of progression that includes Thoreau, Darwin, Marsh, and Muir, as well as Albert Schweitzer and the Russian philosopher P. D. Ouspensky. As Roderick Nash has written, Leopold's "originality should not be distorted" (Callicott, *CSCA* 64), but the mere fact that Leopold's land ethic did not spring ex nihilo does not diminish its singularity.

4. Leopold's articles on game conservation in *The Pine Cone* and numerous speaking engagements before sportsmen's and rancher's associations would win him W. T. Hornaday's Permanent Wildlife Protection Fund's gold medal in 1917, as well as the praise of Theodore Roosevelt, who wrote to Leopold in 1917 saying, "I think your platform [is] simply capital."

5. *A Sand County Almanac* was finally accepted by Oxford University Press in 1948. For a detailed explication of the book's prepublication history see Dennis Ribbens's "The Making of *A Sand County Almanac*" in Callicott, ed., *Companion to A Sand County Almanac*.

6. Leopold's wife recalled that Leopold "thought organized religion was all right for many people, but he did not partake of it himself, having left that behind him a long time ago" (Meine 506). His son Luna described his father's religious beliefs as "kind of pantheistic. The organization of the universe was enough to take the place of God, if you like. He certainly didn't believe in a

personal God, as far as I can tell. But the wonders of nature were, of course, objects of admiration and satisfaction to him" (Meine 506–507).

7. Leopold's testimony was instrumental in instituting an "antlerless" deer season in Wisconsin in 1943 in order to trim a deer herd that had suffered widespread starvation attributed to overpopulation. A significant proportion of the public was incensed about this action, calling it "the crime of '43," and castigated Leopold personally in a number of extremely hostile articles such as one in the June 1945 issue of *Save Wisconsin's Deer*, which attacked him for his "Leopoldian egotism," and for insinuating "that he, the great Aldo, places his knowledge above that of any Wisconsin citizen" (1).

8. For example, in "The Deer Swath" he writes: "there are four categories of outdoors men: deer hunters, duck hunters, bird hunters, and non-hunters. . . . The deer hunter habitually watches the next bend; the duck hunter watches the skyline; the bird hunter watches the dog; the non-hunter does not watch" (208).

9. See, for example John Tallmadge's "Anatomy of a Classic" (110–27) in *Companion to A Sand County Almanac* and Roderick Nash, *The Rights of Nature* (11).

Works Cited

Callicott, J. Baird, ed. *Companion to A Sand County Almanac: Interpretive and Critical Essays* (cited as *CSCA*). Madison: University of Wisconsin Press, 1987.

———. *In Defense of the Land Ethic: Essays in Environmental Philosophy* (cited as *IDLE*). Albany: State University of New York Press, 1989.

Dubos, Rene. *A God Within: A Positive Approach to Man's Future as Part of the Natural World*. New York: Charles Scribner's Sons, 1972.

Leopold, Aldo. *The River of the Mother of God and Other Essays by Aldo Leopold* (cited as *AMG*). Susan L. Flader and J. Baird Callicott, eds. Madison: University of Wisconsin Press, 1991.

———. *Round River: From the Journals of Aldo Leopold*. Luna Leopold, ed. 1953. New York: Oxford University Press, 1993.

———. *A Sand County Almanac* (cited as *SCA*). 1949. New York: Oxford University Press, 1966.

Meine, Curt. *Aldo Leopold: His Life and Work*. Madison: University of Wisconsin Press, 1988.

Nash, Roderick. *The Rights of Nature: A History of Environmental Ethics*. Madison: University of Wisconsin Press, 1989.

Regan, Tom. *The Case for Animal Rights*. Berkeley: University of California Press, 1983.

Ribbens, Dennis. "The Making of *A Sand County Almanac*." In *Companion to A Sand County Almanac*, J. Baird Callicott, ed., 91–109. Madison: University of Wisconsin Press, 1987.

Stegner, Wallace. "The Legacy of Aldo Leopold." In *Companion to A Sand County Almanac*, J. Baird Callicott, ed., 233–45. Madison: University of Wisconsin Press, 1987.

Tallmadge, John. "Anatomy of a Classic." In *Companion to A Sand County Almanac*, J. Baird Callicott, ed., 110–27. Madison: University of Wisconsin Press, 1987.

CHAPTER 8

The New Environmentalism and Rachel Carson's *Silent Spring*

> It is not my contention that chemical insecticides must never be used. I do contend that we have put poisonous and biologically potent chemicals indiscriminately into the hands of persons largely or wholly ignorant of their potentials for harm. We have subjected enormous numbers of people to contact with these poisons, without their consent and often without their knowledge. If the Bill of Rights contains no guarantee that a citizen shall be secure against lethal poisons distributed either by private individuals or by public officials, it is surely only because our forefathers, despite their considerable wisdom and foresight, could conceive of no such problem.
> —Rachel Carson, *Silent Spring*, 1962

> [*Silent Spring*] will appeal to those readers who are as uncritical as the author.... These consumers will include the organic gardeners, the antifluoride leaguers, the worshipers of "natural foods," those who cling to the philosophy of a vital principle, and pseudo-scientists and faddists.
> —William J. Darby, *Chemical & Engineering News*, 1 October 1962

David Brower once summarized the reasons for Rachel Carson's effectiveness as a writer by saying that "she did her homework, she minded her English, and she cared" (Brooks 321). While this terse assessment is essentially accurate, like John Burroughs's succinct prescription for turning natural history into literature,[1] it downplays the literary and rhetorical nuances that distinguish great from competent writing. *Silent Spring*'s effect on American environmental politics is unequaled by any other single book, and Carson's literary skills were readily acknowledged by even her most vehement critics, such as Mississippi congressman Jamie Whitten, who described her as "a former wildlife biologist with a magic pen" (12). The environmental reappraisal triggered by *Silent Spring* was so profound that several writers have compared Carson's political impact to that of Harriet Beecher Stowe.[2] The comparison is an apt one, for, like Stowe, Carson had an uncanny understanding of how the political decision-making process worked and the role public opinion played in that process. As a rhetorician, however, Carson far surpasses Stowe, for in *Silent Spring* she not only alerts her readers to the issue at hand, but she simultaneously educates and persuades them. As Stewart Udall writes in *The Quiet Crisis*, *Silent Spring* was "an ecology primer for millions" (200), and its role in the ecological awakening of America is inestimable. In fact, *Silent Spring* was so effective in alerting the public to the ecological dangers of chemical pesticides that it has become the rhetorical archetype of modern environmental literature, to which the numerous succession of books with titles alluding to environmental Armageddon bears witness.

Within this context, two critical aspects of *Silent Spring*'s political and literary impact merit further exploration. First, *Silent Spring* marks a literary and political dividing line between the conservation era and modern environmentalism. As Robert Cameron Mitchell has pointed out, the conservation movement, as exemplified by the progressive policies of Theodore Roosevelt, has been primarily concerned with "first-generation" environmental issues. Most of these directly relate to land use and wildlife protection, such as the preservation of scenic and wilderness areas and endangered species (84–85). The post–World War II era has brought forth a series of "second-generation" environmental issues that are more complex: toxic chemicals, acid rain, nuclear waste, and carbon dioxide emissions, to name just a few. These problems, posing potential long-range threats to human health and the environment at large, are even more difficult to quantify than the conservation issues of the first generation. The second-generation environmental dilemmas have not only changed the course of the scientific and political debate over the environment, but have also changed the way in which "nature writers"—the term begins to lose much of its usefulness in the context of second-generation issues—contribute to this discourse. As Mitchell has pointed out, most second-generation environmental issues are indirect, difficult to prove, and require a great deal of "scientific detective work." Even then, owing to factual uncertainties, it is often difficult to resolve a controversy to anyone's complete satisfaction (85).

Second, in addition to initiating widespread consideration of an important new set of environmental issues, *Silent Spring* signaled a dramatic change in the tone of environmental discourse. The controversy surrounding the book's publication was a harbinger of the "negative campaigning" that seems invariably to accompany environmental debate (and political debate generally) in the modern era. Carson's success in influencing public opinion and the political process triggered a strong reaction from those who saw her work as a threat either to their economic well-being or to recent gains in agricultural production and other areas in which chemical pesticides played a role. Many of the attacks on *Silent Spring*—particularly by those with vested interests—ignored the merits of Carson's arguments and attempted instead to negate her influence by willfully misrepresenting her positions (accusing her of advocating an immediate and total ban on pesticides, for example) or by attacking her personally, often in a sexist manner, accusing her of bias, hysteria, or worse. While such ad hominem attacks are intellectually dishonest, they can be effective in certain circumstances—primarily in binary, yes/no propositions where the emphasis is on dissuasion ("don't support this person/position") rather than persuasion. However, as Aldo Leopold had noted in "Forestry and Game Conservation" (1918), the debate over conservation issues (let alone that over the more complex second-generation issues) had long since passed from the binary "should we" stage to the more complex "how should we" stage. Thus, *Silent Spring* provides an instructive lesson in the way rhetoric can be effectively used to influence environmental policy, while the criticism it engendered illustrates how the growing tendency, on both sides of the environmental divide, to demonize the other side was destined to impoverish environmental discourse.

Carson was one of the first nature writers to recognize that the battle for public opinion and political influence that had been "won" in the early twentieth century needed to be waged once again, this time for the set of environmental problems represented by second-generation issues. As had been the case in the early days of the conservation movement, the public, their elected officials, and the government bureaucracy had to be alerted to the new problems at hand, educated about the potential hazards, and persuaded to embark upon a remedial course of action. Educating people about the ecological underpinning of these issues was part of the challenge, but, as Carson wrote in *Silent Spring*, there were other factors to contend with before environmental reforms could be implemented, not the least of which dealt with entrenched opposition to the public's right to know:

> This is an era of specialists, each of whom sees his own problem and is unaware of or intolerant of the larger frame into which it fits. It is also an era dominated by industry, in which the right to make a dollar at whatever cost is seldom challenged. When the public protests, confronted with some obvious evidence of damaging results of pesticide applications, it is fed little tranquilizing pills of half truth. . . . It is the public that is being asked to assume the risks that the insect controllers calculate. The public must decide whether it wishes to continue on the present road, and it can do so only when in full possession of the facts. In the words of Jean Rostand, "The obligation to endure gives us the right to know." (13)

While there had been a few federal laws dealing with toxic substances, such as the Federal Insecticide, Fungicide, and Rodenticide Act (1947) and the Pesticide Chemicals Amendment (1954), public attention was far from galvanized, although the potentially deadly side effects of DDT and the related family of chemical pesticides had long been suspected by a number of scientists, including Carson herself, who had unsuccessfully proposed an article on the subject to *Reader's Digest* as early as the mid-1940s. DDT's critical role in helping to control the pestilence spread by the devastations of World War II had made it something of a chemical war hero, resistant to dissenting opinions regarding its widespread use. The effects of DDT and other broad spectrum agents were not limited to "pests" (as Carson points out in *Silent Spring*, they were "biocides," not simply "pesticides"), and their long-term effects on animals, the land, and human health were largely unknown. Still, given the widespread suffering caused by insect-borne disease during the war and the immediate postwar period, such concerns seemed of secondary importance. Although a number of articles were written on the subject of broad spectrum pesticides and their possible undesirable side effects, chemical pesticides were still reaping the public relations bonanza of the war and its ensuing agricultural boom. These articles, some of which appeared in popular magazines such as *Harper's* and the *Atlantic Monthly*, failed to ignite a widespread public reaction, although they may have contributed to a vague and ill-defined fear that the Bomb was not the only new technology to worry about in the atomic age.

Against this backdrop of legislative lethargy in the face of admittedly vague public concerns over toxic chemicals came a series of incidents that led directly to *Silent Spring* and the whole debate over chemical pesticides and their effect on the environment. In the spring of 1957 a group of Long Island, New York, residents led by the well-known ornithologist Robert Cushman Murphy filed suit against the United States Department of Agriculture and other federal and state agencies in Federal District Court, seeking an injunction that would prevent these agencies from following through with a gypsy moth control program that called for the aerial spraying of DDT over large portions of Nassau and Suffolk counties. At the heart of the suit was the plaintiff's contention that such a program was an unconstitutional trespass that would deprive them of property without due process of law in direct contravention of the Fifth and Fourteenth Amendments to the United States Constitution. On May 24, 1957, the District Court denied the plaintiff's motion for a preliminary injunction; while the trial was still pending, the state proceeded with the spraying program.

When the trial (where the plaintiffs sought a permanent injunction against aerial spraying as well as compensation for damages sustained during the 1957 program) finally began in February 1958, the merits of the constitutional issues raised by the plaintiffs were never considered. Instead, the trial judge sided with the state's claim that since the spraying program had already been completed and there were no plans at that time for further spraying, the substantive constitutional issues, as well as the request for an injunction were moot points. As to the issue of damages, although the plaintiffs brought forth evidence at the trial indicating that the chemical spray had caused damage to crops, livestock, fish, and even human health, the judge refused to rule on the issue of whether such harm had in fact occurred, finding only that "the spraying program . . . at the rate of one pound of DDT per gallon of solvent per acre, is not injurious to human health" (164 F. Supp. 120). He then dismissed the case, stating that there was no proof of damages or indication that further aerial spraying of DDT was likely. The Second Circuit Court of Appeals upheld the decision, holding that "the mere possibility" that the spraying program might resume at a future date "is not sufficient to prevent a dismissal for mootness" (270 F.2d 420). The court also upheld the District Court judge's finding that the plaintiffs had failed to prove damages, although this holding did not address the issue of whether damages were in fact suffered, but rather the plaintiff's failure to offer "testimony, expert or otherwise, as to the amount of damages suffered by the plaintiffs" (270 F.2d 421). In fact, in an interesting footnote to the mootness issue, the court noted that, given the "inconvenience and damage as widespread as this 1957 spraying appears to have caused," it would be advisable for courts faced with similar claims to "inquire closely into the methods and safeguards" proposed for such programs.

That same summer, the state of Massachusetts used airplanes to release a mist of DDT mixed with fuel oil over Plymouth and Barnstable counties in an effort to control the mosquito population in that region. One of the places sprayed with the mixture was a private bird sanctuary maintained by Olga Huckins and her husband in Duxbury. In a letter to the Boston *Herald* the following January, Huckins described what ensued at her property after the spraying:

> The "harmless" shower bath killed seven of our lovely songbirds outright. We picked up three dead bodies the next morning right by the door. They were birds that had lived close to us, trusted us, and built their nests in our trees year after year. The next day three were scattered around the bird bath. (I had emptied it and scrubbed it after the spraying but YOU CAN NEVER KILL DDT.) On the following day one robin dropped suddenly from a branch in our woods. We were too heartsick to hunt for other corpses. All of these birds died horribly, and in the same way. Their bills were gaping open, and their splayed claws were drawn up to their breasts in agony. . . . Air spraying where it is not needed or wanted is inhuman, undemocratic, and probably unconstitutional. For those of us who stand helplessly on the tortured earth, it is intolerable. (Brooks 232)

Frustrated in her efforts to stop government agencies from continuing the spraying program in Massachusetts without first studying the immediate and long-term effects of these chemicals on wildlife and humans, Huckins sent a copy of her letter to her friend Rachel Carson, asking her if she had any contacts in Washington who might be able to help. As Carson later wrote in the acknowledgments to *Silent Spring*, this letter "brought my attention sharply back to a problem with which I had long been concerned. I then realized I must write this book."

Rachel Carson was perhaps uniquely suited to put the public's ill-defined uneasiness over toxic chemicals into a book that could appeal to a popular audience without sacrificing scientific accuracy. Her background as a scientist, including a master's degree in marine zoology (Johns Hopkins, 1932) and over fifteen years of experience as a government biologist in the U.S. Bureau of Fisheries (incorporated later in the Fish and Wildlife Service), gave her an expertise that proved immensely helpful in working through the enormous quantity of technical literature on the effects of chemical pesticides. Perhaps even more important to the task of placing this information in an overall ecological context and building a persuasive case for more rigorous controls of such substances were her immense literary talents.[3] By the time she started work on *Silent Spring*, Carson was already a world-reknowned author responsible for three best-selling books: *Under the Sea-Wind* (1941), *The Sea Around Us* (1951), and *The Edge of the Sea* (1955). *The Sea Around Us* stayed on the best-seller lists for nearly two years and won numerous honors, including the John Burroughs Medal for the outstanding book in natural history and the National Book Award, both in 1951. These earlier works foreshadow many of the rhetorical skills that would prove to be vital to the success of *Silent Spring*. As biographer Mary McCay notes, Carson was constantly aware of her audience and how she could best present her scientific material to a nonscientific audience (29–30). Her notebooks for *Under the Sea-Wind*, for example, are full of reminders to herself to present her scientific material accurately without diminishing the sense of identification and drama that is vital to engaging a reader's interest in the narrative.

Carson signed a contract with Houghton-Mifflin to write *Silent Spring* (the working title was *The Control of Nature*) in May 1958.[4] The *New Yorker* acquired the rights to serialize portions of the book prior to its publication date, a technique that Carson knew from her previous publishing experience would ensure a larger audience for the book. Carson had initially hoped to have the book completed by early in 1959, but illness, family responsibilities, and the sheer amount of research and writing involved in such a work pushed the publication date back several years—it wasn't until 1962 that the book was finally completed. As Carson wrote in a letter to Paul Brooks, her editor at Houghton-Mifflin, the delay had its benefits, since a more hasty job might have resulted in a book that was "half-baked, at best." Now, after "long and thorough preparation," Carson was satisfied that, "what I shall now be able to achieve is a synthesis of widely scattered facts, that have not heretofore been considered in relation to each other. It is now possible to build up, step by step, a really damning case against the use of these chemicals as they are now inflicted upon us" (Brooks 244).

As it turned out, the delays in the completion of *Silent Spring* were probably fortuitous, because during the intervening period several events brought the subject of chemical pesticides before the public. In 1959, widespread public concern over cranberries treated with aminotriazol resulted in the Food and Drug Administration (FDA) banning the sale of cranberries sprayed with this pesticide. Less well publicized, perhaps, but just as significant was the long awaited final outcome of the case involving aerial spraying on Long Island. In May 1960, the plaintiffs in *Murphy v. Butler* were denied certiorari by the United States Supreme Court. In a somewhat unusual departure from judicial custom, Justice William O. Douglas filed a lengthy dissent from the court's denial, stating that given the "great public importance" of the issues involved in the case he felt that it was necessary to enumerate the reasoning behind his dissent. Part of this public importance, he wrote, was "increasing concern in many quarters about the wisdom of the use of [DDT] and other insecticides" (362 US 929). Carson had been reluctant to tip her hand publicly while the Murphy case made its way through the court system (although she continued to share new information with the plaintiffs, particularly her friend Marjorie Spock), but she had written a letter on the subject of bird deaths caused by insecticides to the *Washington Post* (April 10, 1959). In his dissent, Douglas quoted this letter at length, concluding with a paragraph that contained the central image—and unanswered question—of *Silent Spring*: " 'To many of us, this sudden silencing of the song of birds, this obliteration of the color and beauty and interest of bird life, is sufficient cause for sharp regret. To those who have never known such rewarding enjoyment of nature, there should yet remain a nagging and insistent question: If this 'rain of death' has produced so disastrous an effect on birds, what of other lives, including our own?' " (362 US 930).

Disappointing as the outcome of the Long Island pesticide case was, Carson had suspected all along that the only real hope for a change in public policy lay with the public. Like George Perkins Marsh and Aldo Leopold, Carson had a Jeffersonian faith that if the people had the facts laid out before them they would make the right decisions and prod their elected representatives to action, and it was to this end that *Silent Spring* was directed. In June 1962, the *New Yorker* ran a three-part serialization of *Silent Spring* that set off a remarkable public debate over chemical pesticides and—just as Carson and her publisher had expected—created a heightened sense of anticipation over the publication of the book, which appeared later that fall. The initial reaction to *Silent Spring* was generally positive—the intense backlash against Carson would not come for several months. The *New York Times* featured *Silent Spring* on the front page of the *Book Review* for September 23, 1962, concluding that "it is high time for people to know about these rapid changes in their environment, and to take an effective part in the battle that may shape the future of all life on earth" (26). A number of respected scientists provided additional weight to Carson's position; for instance, in reviewing *Silent Spring* for the *Saturday Review*, Loren Eisely described the book as "a devastating, heavily documented, relentless attack upon human carelessness, greed, and irresponsibility" (18). *Silent Spring* remained on the best-seller list for thirty-one weeks, selling over a half-million copies in hardcover. For perhaps the first time, the increasingly important medium of television took note of an environmental issue, with CBS airing a special entitled "The Silent Spring of Rachel Carson" in April 1963. No book about nature had ever received such widespread publicity, and Carson's rhetorical skill in describing the potential dangers of chemical pesticides stirred up public opinion as no other writer on the environment ever has, before or since.

Turning to an examination of the rhetorical style of *Silent Spring*, it has been said that Carson had written "a brief of which any attorney might be proud" (Graham 63). While there are a number of similarities between Carson's rhetoric and that used in a legal brief, an even closer analogy may be drawn between *Silent Spring* and a different type of legal argument—that presented by an attorney at trial. As an attorney does in a jury trial, Carson directed her argument primarily to a lay audience, the jury of public opinion, to use a well-worn phrase. At the same time, she had to present the scientific portion of her argument in a way that would satisfy the standards of scientific discourse set by professionals in her field. In other words, the scientists who would assess her arguments in reviews and other commentary on her work would fulfill the role of the judge at a trial, whose task is to make sure that the technical strictures—either of science or of the law—are complied with. The question of Carson's choice of language in *Silent Spring* is clarified when this bifurcated audience is taken into account. Just as a lawyer would use different language when writing for a judge or peer (as with a legal memo or appellate brief) than when speaking to a jury (whose members might be distracted or confused by excessive use of legal jargon), so too did Carson need to speak precisely without resorting to the technical language that might well lose a nonprofessional audience. As Carson herself framed the issue, "my relation to technical scientific writing has been that of one who understands the language but does not use it" (Brooks 3). Likewise, Carson's use of experts and scientific evidence is closely analogous to the way an attorney might use such resources at trial, and is subject to the same strengths and weaknesses. Carson relentlessly built her case up by referring to every bit of evidence—anecdotal, statistical, citations from experts—that she could find, but she accomplished this without interrupting the flow of her argument. Rather than loading down her text with footnotes and citations, she chose instead to supplement it with an enormous appendix (over fifty pages in length) listing her sources. In this way, her citations could be checked—cross-examined, if you will—but the narrative flow of her argument remained intact. Again, as is the case with an attorney at trial who uses witnesses and expert testimony, Carson's argument is as strong as the evidence she used; and, with few exceptions, her evidence proved to be compelling indeed.

When we view Carson's rhetorical style as like that of a trial attorney, we are able to account for two of the most common criticisms leveled against her work—that it is biased and "hysterical." Many reviewers noted that Carson had only presented the evidence *against* chemical pesticides in *Silent Spring*. Some have cited this as evidence of bias; others, such as LaMont C. Cole in his thoughtful and balanced review in *Scientific American*, recognized that the "pro" side of the debate had already "been impressed on the public by skilled professional molders of public opinion" (173). To present both sides of the argument evenhandedly might be appropriate in an academic or purely intellectual context, but to do so in a trial would be malpractice, a disservice to one's client. Carson realized (even if some of her critics did not) that she was serving as the advocate for a "client"—the public that was unknowingly being exposed to potentially dangerous chemicals—and that she must make a forceful argument for its position. Likewise, because her purpose in writing *Silent Spring* was to alert the public to the dangers of chemical pesticides, the case against pesticides had to be a compelling one, both in a scientific and in a narrative sense. Carson had to take the role of an aggressive and flamboyant advocate who is as skilled in the press conference as in the courtroom. The question of her supposed "hysteria" or over-emotionalism involves a fundamental misapprehension of the prosecutorial role she necessarily undertook in this debate. "A Fable for Tomorrow," her opening chapter, was frequently held up by critics as a departure from rational consideration of the facts to indulge in fabricated fantasy about the effect of chemical pesticides on an imaginary community. But in a sense this chapter was not part of Carson's evidence against pesticides at all; it was more like a prosecutor's opening statement, an opportunity for Carson to set the tone of her argument without the evidentiary strictures required in the rest of her presentation. If the parable was alarmist in tone and somewhat overblown—though subsequent incidents such as those at Love Canal and Bhopal indicate that it may not have been so after all—this was because it was a rare opportunity to speculate freely about the effect of chemical pesticides.

There is little doubt that Carson, by assuming the role of prosecutor, anticipated a strong and hostile response from the chemical pesticides industry that might well include attempts to discredit her personally. The industry's ad hominem (or ad feminem, if you will) counterattack began even before publication of the book, when the Velsicol Chemical Corporation attempted to dissuade Houghton Mifflin from proceeding with their plans to publish *Silent Spring*. In a letter that included allegations about "inaccurate and disparaging statements" contained in the book, Velsicol made a thinly veiled suggestion that Carson was either a communist or an unwitting tool of the communists:

Unfortunately, in addition to the sincere opinions by natural food faddists, Audubon groups and others, members of the chemical industry in this country and in western Europe must deal with sinister influences, whose attacks on the chemical industry have a dual purpose: (1) to create the false impression that all business is grasping and immoral, and (2) to reduce the use of agricultural chemicals in this country and in the countries of western Europe, so that our supply of food will be reduced to east-curtain parity. Many innocent groups are financed and led into attacks on the chemical industry by these sinister parties. (Brooks 49)

It was certainly not new for an advocate of environmental reform to suffer personal attacks and innuendo from those with a vested economic interest in the practices targeted by the reform. In the battle to save the Yosemite from development, for instance, John Muir's opponents had seized upon his early days as a shepherd and sawyer in the valley in an attempt to paint him as a hypocrite. However, the very success of *Silent Spring* in generating calls for action against chemical pesticides elicited from the chemical industry and its supporters a response unprecedented in its virulence.

The disparagement of Carson and *Silent Spring* was often accompanied by sexist remarks that played on stereotypes such as "hysterical," "over-emotional," or "illogical." In a December 5, 1962, speech to the Women's National Press Club Carson herself succinctly characterized the campaign against her, saying "the masters of invective and insinuation have been busy: I am a 'bird lover—a cat lover—a fish lover,' a priestess of nature, a devotee of a mystical cult having to do with laws of the universe which my critics consider themselves immune to" (Brooks 303). Another detractor sneered, "I thought she was a spinster. What's she so worried about genetics for?" (Graham 50). It is important to point out, however, that although many of these attacks betray an orientation we would today describe as sexist (and that was often recognized as such even thirty years ago), Carson was not attacked primarily because she was a woman but because she was an environmentalist who represented a danger to some very powerful economic interests. The fact that she was a woman was seized upon as a convenient tool with which to disparage Carson and her work. It was not, however, effective enough to distract attention from her well-reasoned argument, and may actually have backfired against her opponents.

The debate conducted in the pages of the *Chemical and Engineering News*, an industry publication that might be expected to reflect a negative bias against Carson, is an instructive case in point. The first direct mention of *Silent Spring* in the pages of the *Chemical and Engineering News* was a letter to the editor calling for "immediate and respectful recognition" of Carson's study, and pointing out the "sardonic" coincidence of her articles appearing in the *New Yorker* at the same time that *C&EN* was trumpeting an increase in pesticide sales (Caesar).[5] Over the next several months the editorials and articles in *C&EN* that dealt with the issue of pesticides and *Silent Spring* varied in tone from the even-handed detachment of an editor's note responding to the first letter, which spoke circumspectly of the need for separating Carson's "extremely careful reading of the facts" from "implications of the interpretations," to editorials in which Carson was obliquely accused of sensationalism. "Pesticides on Trial," for instance (July 30, 1962), asserted that "shortages of facts and excesses of emotion" could deprive humanity of the benefits of pesticides, thereby resulting in such nightmarish scenarios as "humans dying of starvation in areas where insects destroy crops . . . or dying of disease where insects carry pathogens" (5). Such heavy-handed hyperbole—*C&EN* itself admitted in the August 13, 1962, issue that Carson had never come out in favor of either a complete ban of pesticides or of restricting their use where no alternatives existed (23)—failed to go unnoticed. One reader commented in a letter to the editor in the August 20 issue that *C&EN*'s editorial "suffers from the very faults it finds in others: shortage of facts and excess of emotion" (Muffat). Two weeks later, another reader directed himself to the charge of emotionalism in Carson's *New Yorker* articles, stating that "the emotional quality of the articles is no higher than that of the editorials," and pointedly contrasting Carson's "temperate discussion" with the "irritable" comments of F. A. Soraci (see note 5) reported in the July 16 issue (Blau 4). The reader concluded with a stern admonition that, "if [Carson] has not told the whole truth, she should be refuted with facts, not with insinuations" (Blau 5).

Considering the potential impact of *Silent Spring* on the chemical pesticide industry, including its dread of "the futile circus arena" of Congressional hearings (July 30, 1962, 5), it is striking how measured most of the early letters, articles, and editorial comments on Carson and her work in *C&EN* were. In particular, many letter writers expressed a desire that the industry respond quickly and appropriately to the growing public call for more careful testing and use of chemical pesticides. To a certain extent the early editorial restraint of *C&EN* might be attributed to a forlorn hope on the part of the editorial board and the industry it reflected that *Silent Spring*, like earlier exposés (such as Murray Bookchin's *Our Synthetic Environment*, which had appeared just six months before *Silent Spring*), would have a short shelf life, and that public furor over the concerns Carson had raised would soon die down.[6] However, as calls for new restrictions on chemical pesticides began to intensify and the specter of Congressional hearings loomed, the chemical industry assembled its response, and reactions to *Silent Spring* grew more strident.

The most vitriolic attack on Carson in the pages of *C&EN*—and the one that brought forth the strongest reaction from readers—was a book review in the October 1, 1962, issue written by William J. Darby, which appeared under the heading, "Silence, Miss Carson." Darby claimed that the only people who would find *Silent Spring* appealing would be those readers who were as "uncritical as the author," including: "organic gardeners, the antifluoride leaguers, the worshipers of 'natural foods,' those who cling to the philosophy of a vital principle, and pseudo-scientists and faddists" (60). After a number of hostile comments about Carson's qualifications and the nature of her research, Darby high-handedly dismissed both *Silent Spring* and those who would read it:

> Unfortunately . . . this book will have wide circulation on one of the standard subscription lists [the-Book-of-the Month Club]. It is doubtful that many readers can bear to wade through its high-pitched sequences of anxieties. It is likely to be perused uncritically, to be regarded by the layman as authoritative (which it is not) and to arouse in him manifestations of anxieties and psychoneuroses exhibited by some of the subjects cited by the author in the chapter "The Human Price." (60)

Darby's lofty disdain for the "layman" speaks volumes about why some of Carson's opponents in the debate were at a loss to influence public opinion in the same manner that she herself had. Darby's comment, perhaps sexist in nature, about Carson's "high-pitched sequences of anxieties" becomes unintentionally humorous when read in conjunction with his own postulation about what the end result of Carson's approach to pesticide use would bring: "the end of all human progress, reversion to a passive social state devoid of technology, scientific medicine, agriculture, sanitation, or education" (60).

Darby's review brought forth the strongest *C&EN* reader response on the pesticide controversy. Letters to the editor were tilted slightly in favor of Carson, which is surprising in light of the fact that *C&EN* is an industry magazine. Even a Darby supporter was forced to admit that "Darby does not completely avoid using the same method of indictment he condemns in Rachel Carson . . . excessive emotional appeal and/or other questionable methods" (Clendenin); and a number of other readers called attention to the fact that the Darby review displayed the very lack of scientific detachment and rationality that he had called Carson to task for. One likened Darby's response to that of a "cigarette company executive when someone asks if smoking causes lung cancer" (Good). Another chastised the reviewer by saying, "intemperate remarks are no substitute for reasoned arguments" (Schneour). Several letter writers pointed out parallels between the pesticide question and the recent thalidomide tragedy. One admonished colleagues that "a little humility on the part of chemists is in order" (Meier). Other readers responded in even more adamant terms, calling the review "irresponsible," "bad-tempered," and "hysterical."

The debate that took place in the pages of the *Chemical and Engineering News* was a microcosm of that played out in public forums throughout the nation. The industry's personal attacks against Carson had little, if any, appreciable influence on how the public responded to the issue of chemical pesticides. Carson's faith in how the public—and through them the government—would react once they were informed about the problem proved to be justified. Public pressure led to a series of Congressional hearings, most notably those chaired by Senator Abraham Ribicoff in 1963. It led also to the appointment of a Presidential Commission to evaluate the dangers posed by chemical pesticides[7] and to the introduction of an enormous amount of local, state, and federal legislation that established new procedures and controls for the use of chemical pesticides. However, the most dramatic indication of how far and how fast *Silent Spring* had shifted the public's perception of chemical pesticides may well have been the response to the huge fishkill that took place in the Mississippi and Atchafalaya Rivers in 1963. Instead of shrugging off the fishkills as part of the normal cycle of nature, as they had with earlier though admittedly less widespread fishkills, many people immediately suspected chemical pesticides to be the culprit, and this appears to have been the case.[8] Thus *Silent Spring*, in alerting the public to the dangers of chemical pesticides, precipitated significant political action and inspired reappraisal of the role of chemical pesticides in damaging the environment. Its most enduring legacy, however, may be that it awakened vast numbers of people to their responsibility for the earth's overall ecological health and educated them about basic ecological principles. As G. Kass-Simon writes, "today we know these principles so well that, as with our own language, we are no longer aware that we had to be taught them" (258).

Notes

1. "Truth of seeing and truth of feeling are the main requisite; add truth of style and the thing is done" ("Straight Seeing and Thinking," *Leaf and Tendril* 123).

2. See, for example, Downs 260, Nash 78, and Fox 292.

3. Carson biographer Carol Gartner has pointed out that most critics have erred by situating Carson only "in the field of science or natural history, rather than literature" (1). As Gartner makes clear in her study, this oversight fails to take into account not just the literary merit of *Silent Spring*, but that of Carson's three previous books on the sea as well—all of which merit serious consideration as literature, not just popular science.

4. Carson was at first reluctant to take on the task of writing about the potential dangers posed by the indiscriminate use of chemical pesticides. She had initially tried to get E. B. White to write about the problem in his column for the *New Yorker*, offering to provide him with the research materials she had acquired to that point, but he suggested instead that she write the article herself, and urged her to contact William Shawn, the *New Yorker*'s editor.

5. In that article, *C&NE* had reported that sales of chemical pesticides had increased in the past year despite "adverse publicity from conservationists." They had also quoted, without comment, the remarks of F. A. Soraci, director of New Jersey's Department of Agriculture, who complained that, "In any large scale pest control program in this area, we are immediately confronted with the objection of a vociferous, misinformed group of nature-balancing, organic-gardening, bird-loving, unreasonable citizenry that has not been convinced of the important place of agricultural chemicals in our economy" ("Pesticide Sales Pick Up," July 2, 1962, 22).

6. In a lead article in the August 13, 1962, issue entitled "Industry Maps Defense to Pesticide Criticisms," *C&EN* trotted out a few industry scientists who took issue with Carson's views, but also reported that the industry was divided over how to respond to her, with many believing that "to engage in a public debate with Miss Carson may only call even more attention to her viewpoints than they might otherwise receive" (25). Others in the industry, however, citing the immense popularity of Carson's earlier books on the sea and her reputation for scientific accuracy, felt it was incumbent upon the industry to respond in order to head off public concerns, which might result in "governments at all levels [pushing] for unnecessarily increased regulation" (25).

7. In its report, "The Use of Pesticides," (Washington, D.C.: Government Printing Office, 1963), the President's Science Advisory Committee, appointed by President Kennedy, essentially vindicated much of what Carson had written in *Silent Spring*.

8. See Frank Graham's *Since Silent Spring* (94–108) for an excellent description of the Mississippi fish kill incident and its aftermath. Ironically, it appears that the company primarily responsible for the contamination that led to the fishkill was the Velsicol Corporation—the same company that had attempted to block publication of *Silent Spring*.

Works Cited

Blau, Edmund J. Letter. *Chemical and Engineering News* 3 (September 1962): 4–5.
Brooks, Paul. *The House of Life: Rachel Carson at Work*. Boston: Houghton Mifflin, 1972.
Burroughs, John. *Leaf and Tendril*. Boston: Houghton Mifflin, 1908.

Caesar, G. V. Letter. *Chemical and Engineering News*, July 16, 1962, 5.
Carson, Rachel. *The Edge of the Sea*. New York: Oxford University Press, 1955.
———. *The Sea Around Us*. New York: Oxford University Press, 1951.
———. *Silent Spring*. Boston: Houghton Mifflin, 1962.
———. *Under the Sea-Wind*. 1941. Oxford University Press, 1952.
Clendenin, J. D. Letter. *Chemical and Engineering News*, November 5, 1962, 4.
Cole, Lamont C. Rev. of *Silent Spring* by Rachel Carson. *Scientific American* (December 1962): 173–80.
Darby, William J. "Silence, Miss Carson." Rev. of *Silent Spring* by Rachel Carson. *Chemical and Engineering News*, October 1, 1962, 60–61.
Downs, Robert B. *Books That Changed America*. New York: Macmillan, 1970.
Eiseley, Loren. "Using a Plague to Fight a Plague." Rev. of *Silent Spring* by Rachel Carson. *Saturday Review*, September 29, 1962, 18–19.
Fox, Stephen. *John Muir and His Legacy: The American Conservation Movement*. Boston: Little, Brown and Company, 1981.
Gartner, Carol B. *Rachel Carson*. New York: Frederick Ungar Publishing Company, 1983.
Graham, Frank, Jr. *Since Silent Spring*. Boston: Houghton Mifflin, 1970.
Kass-Simon, G. "Biology is Destiny." In *Women of Science: Righting the Record*, G. Kass-Simon and Patricia Farnes, eds. Bloomington: Indiana University Press, 1990.
Leopold, Aldo. "Forestry and Game Conservation." In *The River of the Mother of God and Other Essays*, ed. Susan L. Flader and J. Baird Callicott. Madison: University of Wisconsin Press, 1991.
Meier, Frank A. Letter. *Chemical and Engineering News* November 5, 1962, 4.
Milne, Lorus and Margery. "There's Poison All Around Us Now." Rev. of *Silent Spring* by Rachel Carson. *New York Times Book Review*, September 23, 1962, 1.
Mitchell, Robert Cameron. "From Conservation to Environmental Movement: The Development of the Modern Environmental Lobbies." In *Government and Environmental Politics: Essays on Historical Developments Since World War Two*, Michael J. Lacey, ed. Washington, D.C.: Woodrow Wilson Center Press, 1989.
Muffat, Peter C. Letter. *Chemical and Engineering News*, August 20, 1962, 5.
Murphy v. Benson. 151 F. Supp. 786. U.S. Dist. Ct. (E.D.N.Y.). May 24, 1957.
Murphy v. Benson. 164 F. Supp. 120. U.S. Dist. Ct. (E.D.N.Y.). June 1958.
Murphy v. Benson. 270 F. 2d. 420. U.S. Court of Appeals (2d Circuit). October 1959.
Murphy v. Butler. 362 U.S. 929; 4 L. Ed 2d 747; 80 S. Ct. 750. U.S. Supreme Court. March 1960.
Nash, Roderick. *The Rights of Nature*. Madison: University of Wisconsin Press, 1989.
Udall, Stewart. *The Quiet Crisis and the Next Generation*. 1963. Salt Lake City: Gibbs and Smith, 1988.
Whitten, Jamie L. *That We May Live*. Princeton: D. Van Nostrand Company, 1966.

CHAPTER 9

Monkey Wrenching, Environmental Extremism, and the Problematical Edward Abbey

> We have yielded too much too easily. It is time to start shoving cement and iron in the opposite direction before the entire nation, before the whole planet, become one steaming, stinking, overcrowded high-tech ghetto. Open space was the fundamental heritage of America; the freedom of the wilderness may well be the central purpose of our national adventure. . . . We need no more words on the matter. What we need now are heroes. And heroines. About a million of them. One brave deed is worth a thousand books. Sentiment without action is the ruin of the soul.
> —Edward Abbey, *Beyond the Wall*, 1984

Early in 1968 Edward Abbey solicited an assignment from a small literary magazine based in Arizona to interview Joseph Wood Krutch, one of Abbey's favorite authors and, like himself, a chronicler of life in the desert southwest. Many years later Abbey wrote in an essay about the meeting that he had long been impressed by Krutch's *The Desert Year* and *The Voice of the Desert*, whose arguments in favor of the preservation of the desert seemed to him to be models of "calm, reasonable, gentle persuasion" (*OLTP* 181). He contrasted Krutch's capacity for rational thought, even when arguing for an issue about which he felt very strongly, with his own affinity for "the impatient, the radical, the violent." Referring to his "self-defeating tendencies as a propagandist," Abbey succinctly characterized his own polemical approach: "A book, I often thought, was best employed as a kind of paper club to beat people over the head with, to pound them into agreement or insensibility" (*OLTP* 181). As with much of Abbey's writing, it may be misleading to read too much into this somewhat facetious statement; yet there is more than just a trace of insightful self-analysis here. As a polemicist for wilderness preservation, Abbey was an outspoken, insistent, often strident advocate with an ecological sensibility clearly in the tradition of Thoreau, Muir, and Leopold. Books such as *Desert Solitaire* (1968) and *The Monkey Wrench Gang* (1975) attracted millions of enthusiastic readers, and the latter inspired one of the more dramatic developments in recent environmental politics, the direct action strategies of groups such as Earth First! and Greenpeace. Yet Abbey's record as an environmental polemicist is mixed.

The combative tone characteristic of Abbey's authorial voice was calculated to challenge and provoke the reader, and, if a series of antagonistic book reviews and angry letters from the public are any indication, many readers rose to the bait.[1] Even many would-be allies in the environmental movement criticized Abbey, some put off by his nonliberal stances on immigration, gun control, or other social issues, and some because they shied away from the confrontational methods he seemed to be (and, I would argue, was in fact) advocating. Abbey's iconoclastic views—he once referred to himself as a "registered anarchist," a typical Abbey paradox—on a variety of issues, including the environment, often seem inexplicable, even contradictory, and he fed this confusion both in his writings and in interviews, where he covered his ideological tracks by confounding those who sought to fit him into pigeonholes. As Abbey's former colleague Peter Wild writes, the "false trail is a calculated part of his literary technique" (189). Numerous other critics have also commented on his "shiftiness."[2] Although Abbey presented himself as a blunt, straightforward speaker, much of his writing is in fact a complex mixture of personal narrative, journalism, philosophy, natural history, political commentary, and storytelling (a "redneck slumgullion stew," he called it), full of paradox, irony, and humor. In a paradox that Abbey would no doubt have found satisfying, some of the same things that help to make his work so rich in a literary sense often serve to diminish its effectiveness as environmental rhetoric. While paradox and textual ambiguity certainly have their virtues, particularly where academicians and literary critics are concerned, in a rhetorical context they are usually counterproductive, serving as little more than a distraction or perhaps a political escape hatch.

Part of the self-generated confusion surrounding Abbey and his work concerns the issue of whether he should be regarded as a nature writer at all. In the introduction to *The Journey Home* (1977), a collection of his essays, he protests against this classification, stating: "I am not a naturalist. I never was and never will be a naturalist. I'm not even sure what a naturalist is except I'm not one" (xi). In another collection of essays, *Abbey's Road* (1979), he plaintively asks, "why, with five published novels and three volumes of personal history to my credit—or discredit if you prefer—why am I classified by librarians and tagged by reviewers as a 'nature writer'?" (xviii). The answer to that question is rooted in (though by no means limited to) one remarkable book, *Desert Solitaire* (1968). Prior to this account of several seasons spent as a ranger at Arches National Monument in Utah,[3] Abbey had enjoyed some small measure of literary success; but none of his earlier works had an impact even remotely approaching that of *Desert Solitaire*. As Abbey acknowledged in an interview, it was this book more than anything else that steered him to writing about nature:

> I never wanted to be an environmental crusader, an environmental journalist. Then I dashed off that *Desert Solitaire* thing because it was easy to do. All I did was copy out of some journals that I'd kept. It was the first book that I published that had any popularity at all, and at once I was put into the "Western Environmentalist Writer" bag, category, pigeonhole. I haven't tried very hard to get out of it. (Hepworth 40)

Abbey's deprecation of *Desert Solitaire*, along with many of his statements about being a nature writer, may well contain some truth but shouldn't be taken at face value. The large number of critical essays written about "that *Desert Solitaire* thing" indicate that, despite Abbey's disparaging comments, there is more to the book than he acknowledges. Paul Bryant makes the point that *Desert Solitaire* is the pivotal point of Abbey's writing, with all of the lines of thought in his later writing in some way passing through it (*SU* 18). Despite his resistance to being labeled a nature writer or naturalist, Abbey's skill at describing and interpreting the natural world, combined with his moral and ethical opinions on mankind's relations with that world, have made his one of the truly original voices in modern nature writing. In short, while critic Ann Ronald is correct in saying that "readers who see in Abbey's books only the word-pictures of an environmental writer or who hear only the polemics of a wild-country advocate have closed their eyes and ears to half his intent and accomplishment"

(94), any reader who would discount his environmental word-pictures and polemics is likewise ignoring a substantial part of Abbey's literary achievement.

Abbey's work displays both a broad familiarity with the genre of nature writing and considerable insight about how his own work relates to that tradition. He professes to admire Thoreau, Muir, Leopold, Beston, Krutch, and Eiseley, among others, but claims that the "constant level of high thinking" characteristic of their work is beyond him: "Some itch in the lower parts is always dragging me back to mundane earth, down to my own level, among all you other common denominators out there in the howling wilderness we call modern American life" (*JH* xii). Instead, Abbey claims, his own literary idols were people like Celine, Dreiser, and B. Traven—"the unloved." Still, hints of Muir, Leopold, and Krutch can be found scattered throughout his work. While Abbey also admires a number of modern nature writers—particularly Annie Dillard, Edward Hoagland, Wendell Berry, and Peter Matthiessen—he derides the tendency of critics to label these writers (including himself) heirs of Thoreau:

> Edward Hoagland, the Thoreau of Central Park and also Vermont. Krutch and Abbey, Thoreaus of Arizona. Wendell Berry, Thoreau of Kentucky, and Annie Dillard, the Thoreau of Virginia and now, I guess, of Puget Sound. John McPhee, the Thoreau of New Jersey, Alaska, Scotland, and anything else he may choose to investigate. Ann Zwinger, the Thoreau of the Rockies. Peter Matthiessen, the Thoreau of Africa, South America, the Himalayas, and the wide wild sea. And others too numerous to mention. (*AR* xx)

Nonetheless, Wendell Berry writes in "A Few Words in Favor of Edward Abbey," despite the tendency of critics to "compare every writer who has been as far out of the house as the mailbox to Thoreau" (6), there are a number of similarities between Abbey and Thoreau, including their environmental sensibility, their lack of political activism, and a pronounced uncongeniality to "group spirit" (6). In Abbey's remarkable essay "Down the River with Henry Thoreau," which intersperses perceptive comments on Thoreau and his work with a narrative about a trip down the Green River in southeast Utah, Abbey confesses that "Thoreau's mind has been haunting mine for most of my life" (13). Abbey greatly admires Thoreau's work, but the essay also contains criticisms of Thoreau (his "unctuous" sermonizing, for example) and may provide the key to understanding the vehemence with which he resisted categorizations and comparisons—even to Thoreau. Abbey begins the essay with a short passage about seeing a bumper-sticker reading "QUESTION AUTHORITY" that highlights his insight into

Thoreau's central motivation as well as his own: "Thoreau would doubtless have amended that to read 'Always Question Authority.' I would only add the word 'All' before the word 'Authority.' Including, of course, the authority of Henry David himself" (*DTR* 14).

Abbey's protestations of innocence concerning the charge of being a nature writer have more to them than simply a desire to be freed from literary categorization. The one consistent theme in his writing that rivals, and to some extent subsumes, his thoughts on wilderness and its preservation, is that of freedom. As a graduate student at the University of New Mexico Abbey wrote his master's thesis on anarchism ("Anarchy and the Morality of Violence"), and in later years he referred to himself as an anarchist, at least in the "ideal realm" (Hepworth 129). At the core of all his writings, including such early novels as *Jonathan Troy* (1954), *The Brave Cowboy* (1956), and *Fire on the Mountain* (1962), as well as later works including *The Monkey Wrench Gang* (1975), *Good News* (1980), and *Hayduke Lives!* (1990), is the idea of personal liberty under siege. As early as 1951, Abbey had identified this theme as the one that most intrigued him, writing in his journal, "the harried anarchist, a wounded wolf, struggling toward the green hills, or the black-white alpine mountains, or the purple-golden desert range and liberty. Will he make it? Or will the FBI shoot him down on the very threshold of wilderness and freedom?" (*CB* 10).

The connection between wilderness and freedom was not only a theme that Abbey explored many times in his fiction, but one that frequently surfaces in his essays as well. It forms much of the philosophical underpinning of his environmentalism and helps to explain why his views often differ widely from those of other advocates of environmental reform. Although many of Abbey's arguments in support of wilderness preservation sound familiar, citing esthetic, utilitarian, spiritual, and ethical rationales, he also has another, "political" argument for preserving the wilderness:

We may need [wilderness] someday not only as a refuge from excessive industrialism but also as a refuge from authoritarian government, from political oppression. Grand Canyon, Big Bend, Yellowstone and the High Sierras may be required to function as bases for guerilla warfare against tyranny. What reason have we Americans to think that our society will necessarily escape the world-wide drift toward the totalitarian organization of men and institutions? (*DS* 130)

This near-apocalyptic vision of totalitarianism recurs frequently in Abbey's writing, both in his fiction (where it reaches a head in *Good News*) and in his essays.[4] While in some ways such a dark vision of American government may recall that of survivalists, camouflage-clad militia members, and other pseudo-revolutionaries, when read in another light, perhaps Abbey's fears seem less far-fetched. After all, he had already fought in one war against totalitarianism (World War II), and even an American president, Eisenhower, had publicly expressed misgivings about the influence of the "military-industrial complex." Rumors, later proven to be true, about the FBI keeping files on hundreds of thousands of American citizens (including Abbey himself) reinforced Abbey's misgivings about the future prospects of democracy in America. As he wrote in "The Second Rape of the West," "There is something in the juxtaposition of big business, big military, and big technology that always rouses my most paranoid nightmares, visions of the technological superstate, the Pentagon's latest fascism, IBM's laboratory torture chambers, the absolute computerized fusion-powered global tyranny of the twenty-first century" (*JH* 180–81).

Abbey's belief that wilderness serves as a protective hedge against the coercive power of the techno-industrial state helps to account for his uncompromising views on wilderness preservation. The question of just how uncompromising, how "extreme" those views were has long been debated, with some, particularly earlier commentators such as Garth McCann and Ann Ronald, claiming that he was essentially a moderate and others, such as Donald Scheese and Peter Wild, characterizing his environmental views as radical.[5] The matter is not easily resolved given the contrarian, paradoxical nature of much of Abbey's writing and the narrative persona he created for himself: the radical environmentalist roaring down the highway in a Detroit land yacht with a pink flower on the hood and an NRA bumpersticker on the back, throwing beer cans out the window while giving a tirade about the destruction of the American wilderness. To some extent, as Abbey himself stated in an interview, this character is "just another fictional creation," (Hepworth 44), but the suggestion of critic Ann Ronald that this persona is merely a radical gesture, a "safety valve for the author's anger and disgust" (72), seems unnecessarily apologetic for Abbey's environmental extremism. Wendell Berry writes in "A Few Words in Favor of Edward Abbey":

[Abbey] is a problem, apparently, even to some of his defenders, who have an uncontrollable itch to apologize for him: "Well, he did *say* that. But we mustn't take him altogether seriously. He is only trying to shock us into paying attention." Don't we all remember, from our freshman English class, how important it is to get the reader's attention? (1)

Although it is possible to explain away Abbey's more radical statements and paint him as a moderate by arguing that one can't take what he says seriously—even when he says, "I am—really am—an extremist" (*JH* xiv)—this forces one to argue that Abbey's essays are little more than extended versions of the "all of my statements are lies" conundrum. Instead, Abbey's radical environmentalism is appropriately situated within the subversive tradition of nature writers such as Thoreau and Muir, and, as Don Scheese has written, Abbey "is the most radical, iconoclastic figure of the lot" (212).

A far more fruitful exercise than attempting to explain away Abbey's radical environmentalism is the challenging task of sifting through his writing for elements of a rhetorically and philosophically consistent environmental ethos. In other words, if Abbey is indeed a radical (albeit one resistant to easy labeling as such), then what are the "radical" elements of his environmental philosophy? Some of his ideas are actually quite moderate, such as the steps he believes necessary to deal with the ills of modern industrial democracy:

Without necessarily rejecting either science or technology, it seems to me that we can keep them as servants not masters. . . . This means, especially in America, defending the family farm against the mechanized monoculture of agribusiness; defending the family ranch against the strip mining company; defending the selective cutting of sustained-yield forestry from the clear-cutting of quick-profit wood products corporations; defending the small town against the spreading BLOB of suburbia; protecting our surviving rivers from the dam-building mania of the politicians; saving our hills and fields, mountains and deserts, roadless areas and wild areas from the aggrandizement of the extractive industries." (*DTR* 118)

Such passages echo Jefferson's ideal of a loosely centered, agrarian democracy, but Abbey isn't merely indulging in nostalgic pastoralism or fighting a rearguard action against "progress." Rather, he aggressively advocates bold measures that would actually increase the amount and the "wildness" of wilderness areas, suggesting, for instance, that the national parks be closed to motor vehicles (*JH* 53, 144). With perhaps just a trace of facetiousness he writes: "About 98 per cent of the land surface of the contiguous USA already belongs to heavy metal and heavy equipment. Let us save the 2 per cent—that saving remnant. Or better yet, expand, recover and reclaim much more of the original American wilderness. About 50 per cent would be a fair and reasonable compromise" (*BTW* xv).

A far more "radical" element of Abbey's environmental philosophy is his rejection of anthropocentrism. In "A Walk in the Desert Hills," he flatly states, "the man-centered view of the world is anti-Christian, anti-Buddhist, antinature, antilife and—antihuman" (*BTW* 40). Abbey makes his first strong declaration against anthropocentrism in *Desert Solitaire*, dramatically (if paradoxically) proclaiming, "I prefer not to kill animals. I'm a humanist; I'd rather kill a *man* than a snake" (17).[6] Perhaps reflecting his interest and academic training in philosophy, Abbey's biocentrism has more in common with the ethics-based response of Thoreau and Muir than with the ecology-based response of Leopold. Although Abbey's statements on biocentric ethics never reach the specificity of such works as Christopher Stone's essay "Do Trees Have Standing?" or Leopold's land ethic, he is adamant that "Human needs do *not* take precedence over other forms of life" (*BTW* 4). He even suggests that, when human consciousness has advanced sufficiently, human ethics may also progress to the point where there might be a "recognition, even, of the right of nonliving things—boulders, for example, or an entire mountain—to be left in peace, alone, for a few centuries now and then" (*DTR* 119). These rights certainly apply to the "animals trapped in our food and research apparatus" (*OLTP* 170–71), and perhaps even to plants. Referring to a "conscious presence" he felt while in the woods near his fire lookout tower on the Grand Canyon's north rim, Abbey (like Muir in *A Thousand-Mile Walk to the Gulf*) speculates about the secret life of trees, "a haunted world where all answers lead only to more mysteries" (*JH* 207). While conceding that a certain amount of logging is needed to satisfy the "reasonable needs of a reasonable number of people on a finite continent," Abbey avers that "Man the Pest" has now reached the point where the few remaining forests are being "massacred." As for me, he writes, "unless the need were urgent, I could no more sink the blade of an ax into the tissues of a living tree than I could drive it into the flesh of a fellow human" (*JH* 208).

The most well-known component of Abbey's environmentalism is also the most radical: his advocacy of direct action—"monkey wrenching" or "ecotage"—aimed against the destroyers of wilderness. "You don't argue with engineers—you have to derail them," Abbey writes in "The Second Rape of the West" (*JH* 169), aptly summing up the difference between mainstream environmental activism and monkey wrenching. A few other radical environmentalists, most notably Murray Bookchin, had argued that the reform measures pushed by mainstream environmentalists fell well short of the radical changes needed, but Abbey is the writer most closely identified with direct action.[7] The concept of ecological sabotage, however, goes back at least as far as Thoreau, who fantasized taking a crowbar to the Billerica dam in *A Week on the Concord and Merrimack Rivers* (see chapter 2). The roots of ecotage might even be traced back to England's Luddites, as Abbey suggests by including in *The Monkey Wrench Gang* a memorial to the eighteenth-century industrial saboteur Ned Ludd. It is perhaps no coincidence that the impetus for Abbey's ideas concerning monkey wrenching was also a dam, in this case the Glen Canyon dam near the border of Arizona and Utah. The flooding of Glen Canyon, where Abbey had rafted and explored prior to the building of the dam, was a radicalizing experience for him, analogous in many ways to Muir's response to the destruction of the Hetch-Hetchy valley. Ironically, in the case of Glen Canyon, environmental organizations such as the Sierra Club not only failed to protect it, but actually bargained it away in order to protect another wilderness area. Compromises like the one that doomed the Glen Canyon probably do much to explain Abbey's ambivalence toward such erstwhile environmentalist allies. Abbey fantasizes a number of times about destroying the dam and letting the canyon gradually revert to its natural state, first in a June 1959 entry in his journal (*CB* 152), and later in works such as *Desert Solitaire* (1968), *The Monkey Wrench Gang* (1975), its sequel, *Hayduke Lives!* (1990), and a number of essays.

The four main characters in *The Monkey Wrench Gang* and *Hayduke Lives!*—Doc Sarvis, Bonnie Abbzug, Seldom Seen Smith, and George Washington Hayduke—are drawn together by their mutual antagonism to "progress" in the desert southwest, as exemplified by the roads, bridges, and dams that have invaded the wilderness areas of that region. Together and separately they perform numerous acts of ecotage, cutting down billboards, disabling construction equipment, and demolishing (or attempting to demolish) bridges, all the while fantasizing about their most spectacular project—blowing up Glen Canyon dam—and eluding frenzied pursuit by those who would put an end to their monkey wrenching. Despite the serious nature of the subject matter, *The Monkey Wrench Gang* is raucously funny, reflecting Abbey's desire to "entertain my friends and exasperate our enemies," as he wrote elsewhere (*VCW* 65). The profane "agent prevaricator," and chief monkeywrencher George Hayduke, who favors "outrageous industrial sabotage" and doesn't shy away from violent confrontation, must continually be reined in by the others, who favor "subtle, sophisticated harassment techniques" (70). As Doc Sarvis cautions Hayduke, "If constructive vandalism turns destructive, what then? Perhaps we'll be doing more harm than good. There are some who say if you attack the system you only make it stronger," to which Hayduke responds, "Yeah—and if you don't attack it, it strip-mines the mountains, dams all the rivers, paves over the desert and puts you in jail anyway" (104).

It is this debate over tactics, the question of whether violence used in the defense of the earth is justifiable, that raises some of the most troublesome issues for environmental activism. In *The Monkey Wrench Gang* Abbey is careful to draw a distinction between violence done to machines, which is morally justifiable to the ecological saboteur, and violence to humans, which is far more difficult to justify. To even get to this point one would have to argue that wilderness protection is a moral imperative that overshadows other ethical considerations (such as legal and ethical strictures against the use of violence), and that mainstream political tactics have failed to protect the wilderness and are unlikely to be successful in the future. Once the ethical issue has been resolved in favor of wilderness protection and mainstream reform has been abandoned as a political tactic, the use of violence as a means of protecting the wilderness may, at least to some, become justifiable. The debate over the use of violence in Abbey's work is particularly intriguing given his longstanding interest in the issue, dating back to his master's thesis on anarchy and the morality of violence. Opinions may even differ among monkey wrenchers, as exemplified by the dialogue about violence, primarily between Hayduke and Doc Sarvis, that continues throughout both *The Monkey Wrench Gang* and *Hayduke Lives!* Asked to recall the first rule of the eco-warrior, Hayduke replies, "Don't get caught," to which Doc Sarvis replies, "No. . . . Rule Number One is, Nobody gets hurt. Nobody. Not even yourself" (*HL* 110). This rule is broken in *Hayduke Lives!*, which may indicate, as Don Scheese has suggested, that Abbey's views on wilderness protection were becoming more extreme late in his life (223-24). His journal tends to corroborate this reading of his fiction as well, advocating "confrontation" and "direct resistance" against developers (344), a tactic that Abbey knew would escalate tensions—as shown in *Hayduke Lives!*—and might well precipitate violent actions by each side.

Some have argued that not only does Abbey not sanction the use of violence, but he isn't advocating the use of ecotage at all.[8] *The Monkey Wrench Gang* and *Hayduke Lives!* are, after all, works of fiction, and to identify the sentiments of a character such as Hayduke with those of Abbey himself is a risky business to say the least. When Abbey's essays and other statements on the issue are taken into account, however, his position on monkey wrenching and ecological sabotage becomes less equivocal. References to monkey wrenching frequently surface in Abbey's essays: pulling out survey stakes (53), chainsawing billboards (219), and musing about "some unknown hero with a rucksack full of dynamite" blowing up the Glen Canyon Dam (165) in *Desert Solitaire*; and advice on desert etiquette in "The Great American Desert" that includes an admonition to "Always remove and destroy survey stakes, flagging, advertising signboards, mining claim markers, animal traps, poisoned bait, seismic exploration geophones, and other such artifacts of industrialism," to name just a few examples. As Abbey wrote in a journal entry in 1979, "It is no longer sufficient to *describe* the world of nature. The point is to *defend* it" (CB 264).

Certainly many readers, including Dave Foreman and other members of Earth First!, a group inspired by *The Monkey Wrench Gang*, took Abbey at his word. Indeed, the activities of eco-raiders inspired by the novel seem to have been predicted by Abbey, who wrote in an epigraph to the book: "This book, though fictional in form, is based strictly on historical fact. Everything in it is real and actually happened. And it all began just one year from today." Earth First! is sympathetically portrayed in *Hayduke Lives!* (the book's title is itself an allusion to the group's name, of course), and Abbey not only defends their activities both in this work and elsewhere,[9] but actually participated in some. In a symbolic bit of guerilla theatre at Glen Canyon dam in 1981, Abbey and other Earth First! demonstrators unrolled a simulated black plastic "crack" down the face of the dam.[10] He also contributed a "Forward!" to *Eco-Defense: A Field Guide to Monkeywrenching*, edited by Dave Foreman and Bill Haywood, in which he explicitly endorses the book and its message, asserting that we are justified in defending the wilderness as we would our homes, *"by whatever means are necessary"* [my emphasis] (8). Finally, and perhaps most compelling, is the testimony of longtime Abbey friends Jack Loeffler and William Eastlake, who reveal that Abbey had in fact engaged in monkey wrenching himself. In an essay published a few years after Abbey's death Loeffler wrote:

Now that Ed lies far beyond the reach of the statute of limitations, it can be revealed that he did *not* limit his attacks against wilderness rapists to his writing ... he physically destroyed those metal marauders that raze wilderness. He pulled up stakes. He closed roads. He did everything he could think of to thwart the juggernaut of so-called human progress *save one thing*—he never, ever caused harm to another human being. (48)

Loeffler goes on to say that, despite Abbey's distaste for violence, Abbey could "easily foresee a time" when the need for such drastic measures might arise. Even without this qualification, it is clear that Abbey felt (as he wrote in the foreword to *Ecodefense*) that "Representative democracy in the United States has broken down" (8), and that extreme measures were now necessary to save what little was left of the American wilderness.

Despite his considerable skills as a writer, his progressive views on the environment, and his success in inspiring radical groups such as Earth First!, it can be argued that Abbey's impact as a polemicist for nature was negligible—perhaps even, as some have suggested, counterproductive. This is certainly so if we judge his record by the usual standards: influencing environmental reform legislation; attracting widespread public support for environmental causes; or persuading the government to set aside wilderness areas for protection. A comparison to other nature writers who were more effective rhetorically (at least in the conventional sense) quickly reveals some of the reasons for Abbey's mixed legacy. First, unlike Muir or Carson, he often gets too far out in front of his audience, flouting his unconventional views, probably knowing full well that radicalism tends to alienate the American public. Even reformers whose private views *were* radical, such as John Muir, took pains to temper their public proposals and couch their rhetoric in familiar, reassuring language in order not to doom them from the start. Not only does Abbey decline to do this, but he often takes a confrontational tone that seems almost calculated to alienate. As Ann Ronald writes, "when he shouts his message, it is least likely to be taken seriously" (199), and Abbey does a good deal of shouting in his writing. Second, the paradox, contradiction, and ambiguities that contribute to making Abbey so interesting (and sometimes frustrating) as a writer, work against him as a polemicist. Although there is often a subtle consistency to what he says, belying an apparent contradiction, his clarity in describing his political beliefs doesn't approach the stylistic clarity he displays when he describes the desert, for example. Finally, unlike most of the other writers discussed in this study, Abbey doesn't posit as his goal, or even as his hope, the achievement of environmental reform through the democratic political system. If representative democracy has broken down, then it is pointless to try to enact environmental reform through that system; indeed, it is pointless to work from within that system at all.

It is this last point that is most important to understanding Abbey's influence, or lack thereof, as a polemicist for environmental reform. He was not a failed rhetorician—indeed, a number of his persuasive essays, such as "Free Speech: The Cowboy and His Cow" and "The Damnation of a Canyon," are perfectly structured examples of classic rhetoric. If Abbey fails to fully use his rhetorical talents in the cause of progressive environmental reform it is because he's after bigger game, so to speak. Wendell Berry's explanation of why Abbey fails to fit in with mainstream environmentalism clearly illustrates this point. Abbey does not believe, writes Berry, that "our environmental problems are the result of bad policies, bad political decisions, and that, therefore, our salvation lies in winning unbelievers to the right political side." Rather, he believes that the root cause of our environmental problems is not political at all, but cultural: "our country is not being destroyed by bad politics; it is being destroyed by a bad way of life. Bad politics is merely another result" (2–3). As a blueprint for how to cure this "bad way of life," Abbey's work falls short; as a jeremiad on the impossibility of continuing to live this way, it has considerable power. In this Abbey may have as much in common with Arne Naess and other deep ecologists as he has with Dave Foreman and Earth First!. As Bill Devall and George Sessions write in *Deep Ecology*, many environmental reformists "feel trapped in the very political system they criticize," and are forced to use the language of resource economists to justify their proposals (3). Like deep ecologists, who propose an alternative societal structure based on sound ecological principles, Abbey avoids the problem of entrapment within the political system by rejecting it outright, arguing that "what we need in our perishing republic is something different. Something entirely different" (*JH* 188).

Notes

1. Abbey seemed to delight in eliciting a heated (often overheated) response from critics, reviewers, and lay readers alike. He gleefully quoted angry responses to his work in books such as *Down the River* and *Abbey's Road*.
2. See for example, David Copeland Morris, "Celebration and Irony: The Polyphonic Voice of Edward Abbey's *Desert Solitaire*" (25).
3. His second novel, *The Brave Cowboy* (1956), was made into a film (retitled *Lonely Are the Brave*) starring Kirk Douglas.
4. See, for example, "Freedom and Wilderness, Wilderness and Freedom" (*JH* 227–38), where he expands upon the idea of preserving the wilderness as a base from which to fight against political repression.
5. In "Edward Abbey and Environmental Quixoticism," for example, Paul Bryant argues that beneath the seeming extremism expressed in Abbey's work is a realistic moderation. Other critics have attempted to split the difference, with one calling him a "radical conservative" (Twining 18), and another referring to him as a "moderate extremist" (Herndon 98).
6. This statement may seem more problematical when considered in conjunction with another passage in *Desert Solitaire* where Abbey describes killing a rabbit with a rock (33–34). However, like Aldo Leopold, Abbey did not consider hunting to be inconsistent with biocentrism. Instead of guilt, Abbey writes that he feels elation: "No longer do I feel so isolated from the sparse and furtive life around me, a stranger from another world. I have entered into this one" (34).
7. Bookchin (who also used the pseudonym Lewis Herber) explored these concepts in works such as *Our Synthetic Environment* (1962), *Post-Scarcity Anarchism* (1970), and *The Ecology of Freedom* (1982). See Roderick Nash's *The Rights of Nature* (164–66) for a discussion of the impact of Bookchin's work on the environmental movement.

8. Several critics, including Ann Ronald, Garth McCann, and Scott Slovic have argued that Abbey is not advocating illegal action (partly because he knows that it won't work) but is instead trying to alert people to the issue of wilderness preservation, to shock them into awareness. Picking up on Ann Ronald's point that Abbey uses "his sense of humor to pronounce a sobering message," Scott Slovic writes, "Rather than merging together to "pronounce" Abbey's "sobering message" about the environment, the aesthetic and moral currents in *The Monkey Wrench Gang* strain to become separate, like oil and water; they produce a tense disjunction which forces us to stay on our toes" (103).

9. For instance, in a journal entry for September 16, 1987, he writes: "That sloppyminded asshole Alston Chase, in *Outside* (October '87) calls EF! and Sea Shepherd and animal rights groups 'eco-terrorists.' The scum. Must send a nasty letter to *Outside*" (*CB* 335).

10. See Nash, *Rights of Nature* 189–98 for a discussion of this incident and other actions by radical environmentalists.

Works Cited

Abbey, Edward. *Abbey's Road* (cited as *AR*). 1979. New York: Plume, 1991.
———. *Beyond the Wall: Essays from the Outside* (cited as *BTW*). New York: Henry Holt, 1984.
———. *The Brave Cowboy: An Old Tale in a New Time*. New York: Dodd, Mead, 1956.
———. *Confessions of a Barbarian: Selections from the Journals of Edward Abbey, 1951–1989* (cited as *CB*). David Petersen, ed. Boston: Little, Brown, 1994.
———. *Desert Solitaire: A Season in the Wilderness* (cited as *DS*). New York: Simon & Schuster, 1968.
———. *Down the River* (cited as *DTR*). New York: E. P. Dutton, 1982.
———. *Fire on the Mountain*. New York: Dial Press, 1962.
———. *Good News*. New York: Dutton, 1980.
———. *Hayduke Lives!* (cited as *HL*). Boston: Little, Brown, 1990.
———. *Jonathan Troy*. New York: Dodd, Mead, 1954.
———. *The Journey Home: Some Words in Defense of the American West* (cited as *JH*). 1977. New York: Plume, 1991.
———. *The Monkey Wrench Gang*. 1975. New York: Avon Books, 1976.
———. *One Life at a Time, Please* (cited as *OLTP*). New York: Henry Holt, 1987.
———. *A Voice Crying in the Wilderness (Vox Clamantis in Deserto): Notes from a Secret Journal* (cited as *VCW*). New York: St. Martin's Press, 1989.
Berry, Wendell. "A Few Words in Favor of Edward Abbey." In *Resist Much, Obey Little: Some Notes on Edward Abbey*, James Hepworth and Gregory McNamee, eds. 1985. Tucson, Ariz.: Harbinger House, 1989.
Bryant, Paul. "Edward Abbey and Environmental Quixoticism." *Western American Literature* 24:1 (May 1989): 37–43).
———. "The Structure and Unity of *Desert Solitaire* (cited as *SU*)." *Western American Literature* 28:1 (May 1993): 3–19.
Devall, Bill, and George Sessions. *Deep Ecology: Living as if Nature Mattered*. Salt Lake City: Gibbs Smith, 1985.
Eastlake, William. "A Note on Edward Abbey." In *Resist Much, Obey Little: Some*

Notes on Edward Abbey, James Hepworth and Gregory McNamee, eds. 1985. Tucson, Ariz.: Harbinger House, 1989.

Foreman, Dave, and Bill Haywood, eds. *Ecodefense: A Field Guide to Monkeywrenching*. 1985. Tucson, Ariz.: Ned Ludd Books, 1989.

Hepworth, James. "The Poetry Center Interview (1977)." In *Resist Much, Obey Little: Some Notes on Edward Abbey*, James Hepworth and Gregory McNamee, eds. 1985. Tucson, Ariz.: Harbinger House, 1989.

Herndon, Jerry A. " 'Moderate Extremism': Edward Abbey and 'The Moon-Eyed Horse.'" *Western American Literature* 16:2 (August 1981): 97–103.

Loeffler, Jack. "Edward Abbey, Anarchism and the Environment." *Western American Literature* 28:1 (May 1993): 43–49.

McCann, Garth. *Edward Abbey*. Boise State Western Writers Series No. 29. Boise, Id.: Boise State University Press, 1977.

Morris, David Copeland. "Celebration and Irony: The Polyphonic Voice of Edward Abbey's *Desert Solitaire*." *Western American Literature* 28:1 (May 1993): 21–31.

Nash, Roderick. *The Rights of Nature: A History of Environmental Ethics*. Madison: University of Wisconsin Press, 1989.

Ronald, Ann. *The New West of Edward Abbey*. Albuquerque: University of New Mexico Press, 1982.

Scheese, Don. "*Desert Solitaire*: Counter-Friction to the Machine in the Garden." *North Dakota Quarterly* 59:2 (Spring 1991): 211–27.

Slovic, Scott. *Seeking Awareness in American Nature Writing*. Salt Lake City: University of Utah Press, 1992.

Twining, Edward S. "Edward Abbey, American: Another Radical Conservative." *Denver Quarterly* 12:4 (1978): 3–15.

Wild, Peter. *Pioneer Conservationists of Western America*. Missoula: Mountain Press Publishing Company, 1979.

Conclusion

In Steven Lewis Yaffee's fine study of the spotted owl controversy in the Pacific Northwest, *The Wisdom of the Spotted Owl: Policy Lessons for a New Century*, Yaffee succinctly frames the threshold issue that today's environmental reformers must overcome in order to change American environmental policy:

> The decisionmaking system that we have is remarkably stable, providing us with a governance process that is basically the same that has been in effect in the country for more than two hundred years: an extraordinary state of affairs when compared with most other modern-day societies. The process was designed to ensure such stability, allowing for transitions of power and personalities in ways that are not disruptive to ongoing social or economic activities. But as a result, the set of decisionmaking institutions yields decisions that are conservative in nature, resist significant change, and generally seek to perpetuate status quo conditions. In addition, the process is protective of itself, that is, it limits changes in the way choices are made. (200)

In addition to a political system that tends to preserve the status quo, the American people are also conservative in a fundamental way that transcends the cyclical swing to the right that took place in the 1994 elections. Some environmental advocates argue that the present system is so inherently flawed that it needs to be torn down and reconstructed according to an ecologically responsible paradigm. Given the political realities of the situation, however, rapid change of this order seems unlikely, barring an ecological catastrophe so severe that profound political transformation cannot be escaped.

This does not mean that significant, even "radical," change cannot be effected through the present system, merely that change will often be painfully slow and resistant to radical means. To borrow Walt

Whitman's admonition to Horace Traubel, "Be radical, be radical, be radical; be not too damned radical." Given the inherent conservatism of the American people and political system, it is vitally important for environmental writers and advocates not to outdistance their audience. In the area of environmental reform the most politically effective work has been done by writers and advocates who framed their issues carefully, knew their audience, and understood how public opinion interacts with the political system to effect political change. In other words, the polemical skills of writers/advocates such as Muir, Roosevelt, Leopold, and Carson have clearly demonstrated that environmental writing and environmental reform are inextricably linked, and that quite often the first is a prerequisite for the second.

Due in large measure to the work of the nature writers discussed in this study, and many others, the way in which Americans see nature has changed considerably in the last century, and this change has had significant political repercussions, including substantial movement toward ecologically responsible practices and policies.

Where nature writers such as John Burroughs and John Muir once had virtually to create an audience for nature writing and environmental reform out of thin air, there is now an enormous constituency that identifies with the goals of the environmental movement. In a 1994 Los Angeles *Times-Mirror* poll, for instance, 78 percent of those questioned said that the United States government should do whatever it takes to protect the environment. There is even some indication that a number of religious denominations are seeking to reexamine the link between personal responsibility to the environment and traditional biblical teachings.[1] In addition to this strong public sentiment in favor of environmental protection, the explosive growth in the membership of organizations such as the Sierra Club, the Audubon Society, Greenpeace, and similar advocacy groups during the 1970s and 1980s has produced a powerful lobby capable of bringing enormous resources and political pressure to bear in favor of environmental legislation. Elected officials have responded by enacting a vast number of laws in the past thirty years, including landmark legislation such as the National Environmental Protection Act, the Endangered Species Act, the clean air and water acts, and legislation creating the Environmental Protection Agency and the Superfund for the cleanup of toxic waste sites.

Given these successes as well as the changing nature of environmental politics in this country, it might well be asked whether nature writers still have a significant role to play in effecting environmental reform. Nature writers today are less likely to be the immediate force behind the creation of environmental advocacy groups, the passage of legislation, or other reforms than was once the case, and more likely to play the part of political gadfly. The polemical function of nature writers has been diminished as television, radio, and the direct mail of advocacy groups have come to play a greater role in environmental politics. It is also an inescapable if sometimes regrettable fact that environmental reform in the latter half of the twentieth century has become a highly professionalized business, with scientists, government decision makers, industry representatives, and professional lobbyists on both sides taking the lead in decision making. While this is not a new phenomenon—after all, such pioneer conservationists as Gifford Pinchot and Aldo Leopold are early examples of the trend toward professionalization—it has accelerated greatly in the past few decades.[2] Many of today's complex environmental issues require a level of technical expertise that all but precludes the layperson from resolving (or sometimes even fully understanding) them. As the specialists come in, bringing with them the often obscure jargon, methods, and analyses of professional and academic discourse, the number of writers with an ability to speak both this language and that of the general public (as Rachel Carson was able to do) decreases drastically. Even specialists sometimes find it difficult to keep up with the demands that the interdisciplinary nature of environmental policy-making places upon them. As David A. Adams writes in *Renewable Resource Policy*, a text for students and practitioners in the field of resource policy, "No longer does a scientific-technical-administrative background suffice for a natural resource manager. He or she must also have legal-political expertise" (83).

The most significant recent development affecting the role of nature writers in environmental reform, however, may well be the fact that the battle over environmental protection now often takes place in the courtroom. Since the early 1960s, legal theorists such as Clarence Morris, Christopher Stone, and David F. Favre have published a number of groundbreaking articles and treatises questioning the anthropocentric bias of the American legal system and suggesting that all living things are entitled to certain rights. Stone's essay "Should Trees Have Standing?" (1972) has been particularly influential, and was cited with approval by Justice William O. Douglas in his famous dissent to *Sierra Club v. Morton*, 405 U.S. 727 (1972). While that case, in which the court ruled that the Sierra Club did not have standing to sue in order to block the proposed Mineral King development in the Sierras, was a momentary setback to the environmental movement, it represents the first case in which the question of the rights of nonhuman species was considered by the Supreme Court (even if only from a procedural point of view). It also led to the passage of legislation that made it easier for environmental groups to sue to intervene in these cases, legislation that, along with the creation of the Environmental Protection Agency (EPA) in 1970, has helped to make litigation the remedy of choice in many environmental controversies; indeed, the mere threat of EPA action or a civil suit has convinced many a polluter to "voluntarily" curtail harmful practices. As an intricate legal infrastructure of federal and state statutes and regulations, case precedents, and administrative decisions has developed, the legal system has become an increasingly significant forum for environmental discourse. Much of the most creative contemporary thinking in environmental protection is now exhibited in legal briefs, law review articles, and judicial decisions; in short, fifty years from now we may well consider people such as Christopher Stone and William O. Douglas as the most influential "nature writers" of our era.

A number of potential hazards, however, accompany an increased reliance on the legal system to resolve environmental controversies. First, the confrontational nature of litigation may exacerbate polarization, and thereby hinder the development of new approaches to environmental protection that would require a measure of cooperation between regulated interests, the government, and environmental reformers. Examples of this type of approach include a restructuring of the tax system or innovative uses of the free market system.[3] Second, litigation is a two-edged sword; environmental reformers have been so successful in using the legal system to advantage that their opponents have adopted similar tactics in order to slow down or halt reform. For instance, in an attempt to intimidate environmental whistleblowers, some polluters have begun to use lawsuits as a weapon, thereby discouraging all but the most determined (and well financed) groups and individuals from bringing civil suits against them. Third, there is always the possibility that laws and decisions supporting environmental protection may be overturned, particularly in the present political climate. For example, on June 29, 1995, environmentalists won a major victory with the Supreme Court's decision in *Babbitt v. Sweet Home Chapter of Communities for a Greater Oregon* (No. 94-859), but immediately following the court's ruling that the Endangered Species Act barred the destruction of essential habitat for endangered species even on private land, opponents intensified pressure on Congress to rewrite the law. Fourth, once environmental reform is taken out of the realm of public discourse (or if the public perceives it to be so taken) and is placed in the hands of professionals such as attorneys, the possibility increases that environmentalism may be stigmatized by the

Conclusion 171

charge of elitism, a charge that has dogged the movement almost since its inception. Finally, an overreliance on litigation bypasses what may well be the environmental movement's most powerful weapon—public support for environmental reform issues. As Aldo Leopold wrote in "Land Use and Democracy" (1942), "When the Audubon Society killed the millinery feather trade in 1913, what was its real weapon, the prohibitory law or the refusal of intelligent women to buy wild bird plumage? The answer is plain. The law was merely the symbol of a conviction in the mind of a minority" (296).

Some contemporary environmental advocates and writers may themselves be unintentionally diminishing their political contribution by rhetorical excesses and miscalculations. In *Ecological Ethics and Politics*, H. J. McCloskey identifies two of the most common flaws in today's ecological political writing:

> one feature relates to the readiness of many political theorists on the basis of the flimsiest evidence to waive lightly man's claim to recognition of his human rights on the ground that the ecological crisis is so grave and critical that only the most drastic, desperate measures will succeed. The other feature is the very great confidence in the capacity of states to take whatever action is necessary to avoid the ecological crisis. In fact, the states of the world today greatly contribute to various of the ecologically based problems that confront mankind. They are unworthy of the great faith and trust that many ecological reformers appear to have in them or in the state in the abstract. (108)

There are, of course, numerous instances in which ecological crises *have* been grave enough to call for drastic measures, and it is sometimes necessary to resort to overstatement in order to gain public attention, but doomsaying has become such a formulaic response to environmental controversies that polemicists who resort to such a strategy risk inuring the public to such calls and, like the boy who cried wolf, may eventually lose their audience entirely.[4] Gregg Easterbrook, one of several recent critics of environmental doomsaying, argues that such pessimism is unwarranted on the facts and has more to do with fundraising and the internal dynamics of the environmental movement than it does with impending ecological catastrophe (25). While Easterbrook's litany of successes on the environmental front is often overly rosy, his advice to reformers that they acknowledge areas in which progress has been made is sound even if considered merely from the perspective of rhetorical effectiveness. Environmental advocates would enhance their credibility and defuse the oft-repeated charge by opponents that they are being "unreasonable" if they exhibited a semblance of evenhandedness, at least in their presentation of facts and their disputes over facts. The alternative, as Easterbrook writes, is that they risk pushing the public to what he calls the " 'oh, shut up' point" (25).

Illustrative of this point is the increasing frequency with which terms such as "green fatigue" and "ecological overload" have appeared in media stories about hard times in the environmental movement. Despite the fact that there is still a great deal of public support for green issues, environmental organizations have experienced a significant decline in financial support.[5] These phenomena are no doubt attributable at least in part to political cycles and to the lack of any recent high-profile ecological disasters, such as the *Exxon Valdez* incident or the appointments as interior secretary of James Watt and Manuel Lujan. One might also point to the perhaps illusory sense of security that comes with an administration that is presumed to be more sympathetic to the environment than were the Reagan and Bush administrations. However, even some environmentalists have pointed out another possible causative factor—green fatigue stemming from the rhetorical overkill that has become such a staple of environmental advocacy and writing. For example, in an essay entitled "Green Guilt and Ecological Overload" that appeared on the op-ed page of the *New York Times*, historian and longtime environmentalist Theodore Roszak questioned whether "green guilt, the mainstay of the [environmental] movement, has lost its ethical sting" due to its overuse and misuse (June 9, 1992).

All of this comes at a time when forces opposed to environmental reform have renewed their attacks on the laws and regulations enacted over the last thirty years. Beginning in the 1970s, antiregulatory efforts by such groups as the Sagebrush Rebellion, property rights advocates, and the ironically named "wise use" movement have spread and intensified. In many ways, these groups differ little from the vested interests that have fought against reform since the earliest days of the conservation movement. Often they are little more than well-funded attempts on the part of subsidized ranching, mining, and logging interests to limit the federal regulation of public lands. In alliance with sympathetic politicians on the state and federal levels, however, they have already influenced the debate over environmental protection, and will undoubtedly continue to do so in the foreseeable future. Under the banner of "states' rights," these groups have sought—with some success—to identify environmental protection with big government and liberalism, anathema to many, particularly in the west. While it can hardly be claimed that such groups are completely responsible for the growing polarization of environmental discourse, there is little doubt that they have contributed substantially to a political atmosphere in which federal land use personnel find themselves confronting citizens

Conclusion

who are resentful of what they see as unfair restrictions on their right to use either their own or the public lands.

An even more significant development in environmental politics, is the shift in political power brought about by the Republican victory in the 1994 Congressional elections. While the "Contract With America" didn't feature the dismantling of America's environmental protection laws as one of its themes, the attack on "over-regulation" is a thinly veiled attempt to roll back federal environmental regulations that protect the land, air, and water. Since the elections, the Republican leadership—particularly in the House of Representatives—has begun a review of environmental laws such as the Superfund, the Endangered Species Act, the Clean Air Act, and the Clean Water Act that may well result in vastly weaker federal laws and regulations in these areas. Federal agencies charged with the enforcement of existing laws, such as the EPA, the Fish and Wildlife Service, the Forest Service, and the Bureau of Land Management, have also come under political fire, and House Republicans have proposed to cut funding for the National Biological Survey entirely, in large measure due to anger over their research on endangered species.[6] Given strong public support for the environment, these efforts might be expected to backfire politically, but their proponents have thus far been circumspect in presenting their proposals, speaking in terms of job creation, deregulation, or shifting duties to the states, rather than of rolling back environmental protection. An article in the December 12, 1994, issue of *Business Week* aptly characterized the Republican challenge to green laws as a "guerilla war" rather than a "direct attack" (102), but in either case the result would be the same—a weakening of the laws and regulations governing environmental protection.

In large measure these setbacks have come about because environmental advocates are no longer as successful as they once were in setting the terms of the debate over the environment. The importance of winning the rhetorical highground becomes immediately evident when one considers the spotted owl case, where the logging industry and its supporters succeeded, at least in part, in depicting the struggle as one that pitted "owls against jobs." Likewise, "wise use" and property rights advocates have furthered their agenda by adeptly mixing anecdotal evidence of regulatory excess with the populist mythos of self-sufficiency and independence and the general undercurrent of dissatisfaction with the federal government. If nothing else, their inroads should remind environmental advocates that the process of alerting, informing, and persuading the public and its elected representatives regarding environmental issues is a continuing one. As David Helvarg writes in *The War Against the Greens*, "the Wise Use/Property Rights backlash has been a bracing if dangerous reminder to environmentalists that power concedes nothing without a demand and that no social movement, be it ethnic, civil, or environmental, can rest on its past laurels" (458).

Challenges to environmental reform from groups such as these suggest that the need for nature writers to help frame the terms of the political debate over the environment is as vital today as it was in the days of John Muir. In *African Silences*, Peter Mathhiessen describes the factors that led to international curtailment of the ivory trade, writing that the ban was brought about "under pressure from strong public sentiment as well as effective lobbying by conservation groups" (223). This combination of forces is still the key to meaningful reform measures, and it is in helping to form public sentiment that nature writing continues to play an important role in environmental politics. There is a continued need for nature writers with the skill to reach and influence the public, and earlier writers such as those discussed in this study provide us with a number of important lessons in rhetoric and advocacy that are still relevant today.

Perhaps the most important of these lessons is the careful attention to audience demonstrated by these writers. Realizing that not everyone would be swayed by ethical arguments in favor of issues such as wilderness protection or the curtailed use of pesticides, writers such as Muir, Leopold, and Carson all took care not to frame their arguments in strictly moral terms, but to employ a full range of rationales for reform. They also took care not to get too far out in front of their readers. Even John Muir toned down some of his more radical biocentric beliefs when writing to persuade a general audience, a tactic that did not betray his core beliefs, but demonstrated instead a wholly appropriate sense of how far he might be able to bring his audience. The efforts of these writers to expand their audience and reach beyond the committed core of believers is also worthy of note. There is an unfortunate tendency today among environmental writers and advocates to preach to the converted, rather than to attempt the far more difficult task of converting the unconvinced. In this area, the example of Aldo Leopold is particularly instructive, since many of his essays attempted to do just this—to convince farmers, ranchers, hunters, and other land-users that land use reform is in everyone's best interest and not merely something urban conservationists were seeking to foist upon them.

Fortunately for the fate of environmental reform, there may well be more outstanding nature writers today than there have ever been before. Peter Matthiessen, Gary Snyder, John McPhee, Annie Dillard,

Barry Lopez, Wendell Berry, Edward Hoagland, David Quammen, and Tim Cahill, to name just a few, have all helped to move the political debate over the environment in the direction of ethical and ecological responsibility. Nature writing may also be more popular than ever, judging from the numerous specialized forums for ecological literature and the "nature" section contained in virtually every commercial bookstore. This success may be accompanied by a whole new set of challenges, however. As nature writing gathers its own discrete reading audience, the genre is increasingly relegated to specialty magazines such as *Outside, Sierra,* and *Audubon* rather than the great general circulation periodicals. Unlike the days of Burroughs and Muir, there is no great publishing house that promotes the work of its coterie of nature writers by featuring their essays in a house magazine of wide circulation, as Houghton Mifflin once did with the *Atlantic Monthly*. If nature writers are to influence the debate over the environment in a meaningful way, they must continue to expand their audience, to reach out beyond the committed core of sympathizers who are already predisposed to nature writing and environmental reform. It is in the mainstream of American culture that today's nature writers will find the greatest rhetorical challenges and political opportunities, and it is there that the fate of environmental reform will be decided.

Notes

1. See for example, "Religions Are Putting Faith in Environmentalism," *New York Times*, November 6, 1994, 34:1.

2. As early as 1918, in "Forestry and Game Conservation," Aldo Leopold observed that the debate over forest conservation had progressed from the "question of *whether* our forests should be conserved . . . [to] the question of *how* our forests should be conserved," and it was his opinion that it was the class of professional foresters that would answer this second question (55).

3. See, for example, "Cheapest Protection of Nature May Lie in Taxes, Not Laws," *New York Times*, November 24, 1992; and "Environmentalists Try to Move the Markets," *New York Times*, August 22, 1993.

4. A number of recent books, including John W. Maddox's *The Doomsday Syndrome*, Ronald Bailey's *Ecoscam: The False Prophets of Ecological Apocalypse*, Martin W. Lewis's *Green Delusions: An Environmentalist Critique of Radical Environmentalism*, and Gregg Easterbrook's *A Moment on the Earth*, have discussed the tendency of many ecological writers to overindulge in doomsday scenarios. While some critics of ecological doomsayers have their own political axe to grind, their general point is still a valid one.

5. See, for example, "For the Environment, Compassion Fatigue," *New York Times*, November 6, 1994; "Great Green Shakeout," *Outside*, July 1991, 19–20; "Green Groups Enter a Dry Season as Movement Matures," *Wall Street Jour-*

nal October 21, 1994; "Selling Out? Pushed and Pulled, Environment Inc. Is on the Defensive," *New York Times*, March 29, 1992; and "Big Environment Hits a Recession," *New York Times*, January 1, 1995.

6. See "House Panel Agrees to Slash Support for Arts and Humanities," *The Chronicle of Higher Education*, June 30, 1995.

Works Cited

Adams, David A. *Renewable Resource Policy*. Washington, D.C.: Island Press, 1993.

Bailey, Ronald. *Ecoscam: The False Prophets of Ecological Apocalypse*. New York: St. Martin's Press, 1993.

Easterbrook, Gregg. "Green Cassandras." *The New Republic*. July 6, 1992, 23–25.

Hays, Samuel P. "Three Decades of Environmental Politics: The Historical Context." In *Government and Environmental Politics: Essays on Historical Developments Since World War Two*, Michael J. Lacey, ed. Washington, D.C.: The Woodrow Wilson Center Press, 1989.

Helvarg, David. *The War Against the Greens*. San Francisco: Sierra Club Books, 1994.

Leopold, Aldo. *The River of the Mother of God and Other Essays by Aldo Leopold*. Susan L. Flader and J. Baird Callicott, eds. Madison: University of Wisconsin Press, 1991.

Lewis, Martin W. *Green Delusions: An Environmentalist Critique of Radical Environmentalism*. Durham: Duke University Press, 1992.

McCloskey, H. J. *Ecological Ethics and Politics*. Totowa, N.J.: Rowman and Littlefield, 1983.

Maddox, John. *The Doomsday Syndrome*. New York: McGraw Hill, 1972.

Matthiessen, Peter. *African Silences*. New York: Random House, 1991.

Taylor, Paul. *Respect for Nature: A Theory of Environmental Ethics*. Princeton: Princeton University Press, 1986.

Yaffee, Steven Lewis. *The Wisdom of the Spotted Owl: Policy Lessons for a New Century*. Washington, D.C. Island Press, 1994.

Index

Abbey, Edward, 4, 6; as advocate for wilderness preservation, 152–53; connection between wilderness and freedom 156–57; *Desert Solitaire*, 154, 161; environmental philosophy of, 158; *Hayduke Lives!*, 156, 160–62; *The Monkey Wrench Gang*, 156, 159–62; monkey wrenching, 159–61; as nature writer, 154–56; radicalism of, 157–59, 161–64; rejects anthropocentrism, 158–59, 164n; rhetorical effectiveness of, 153, 163; on Thoreau, 155–56
Abolition, 35, 49
Adams, David A., 169
Adirondack Park, 64–65
Anthropocentrism, 1, 3, 30–31, 61
Audubon, John James, 23, 108

Babbitt v. Sweet Home Chapter of Communities for a Greater Oregon, 170
Bade, William F., 90, 92, 96
Ballinger, Richard, 102
Barlow, Joel, 21, 26n
Barrus, Clara, 73, 78–80, 81n
Bartram, John, 18, 25n
Bartram, William, 16–19, 25n
Benton, Myron, 74, 76
Berry, Wendell, 155, 157, 163, 174
Beston, Henry, 41
Beverly, Robert, 18
Bierstadt, Albert, 24, 26n
Biocentrism, 3, 6, 30; compared to abolition, 35

Bookchin, Murray, 147, 159, 164n
Blackstone, William, 16
Boone, Daniel, 23
Boone and Crockett Club, 112–13. *See also* Roosevelt, Theodore
Bounty system, 14, 25n
Bradford, William, 9, 11, 18
Brooks, Paul, 56, 63, 79; as Rachel Carson's editor, 142
Brower, David, 136
Brown, Charles Brockden, 21
Bryant, William Cullen, 21, 49, 64
Bryant, Paul, 154, 164n
Buell, Lawrence, 79
Buffon, Georges de, 17, 20
Burbick, Joan, 47
Burroughs, John, 2, 4, 18, 85, 92, 95, 136, 168; aversion to politics, 78–79; compared to Thoreau, 70, 75; Darwin's influence on, 72–73; as ecologist, 74; on Emerson, 33, 63, 73; Emerson's influence on, 71–72, 81n; friendship with Roosevelt, 78, 118–20 (*see also* Roosevelt, Theodore); on hunting, 78; impact on conservation movement, 70, 78–80; literary reputation of, 69; popularizes nature study, 68–69, 77–78; rejects anthropocentrism, 70, 72; religious beliefs of, 71, 74–76; Whitman's influence on, 72–74, 81n
Byrd, William, 18

Cahill, Tim, 175

Callicott, J. Baird, 124
Carr, Ezra S., 97
Carroll, Peter N., 12
Carson, Rachel, 3, 4, 60, 123, 133, 163, 168, 174; attacks on, 146–49; dividing line between conservation movement and modern environmentalism, 137; effect on environmental politics, 137–138, 149; impact of *Silent Spring* 142–44; incidents leading to publication of *Silent Spring*, 139–41; reaction to *Silent Spring*, 146–49; rhetorical technique of, 144–45
Catesby, Mark, 18
Catlin, George, 24, 26n, 48, 59
Cohen, Michael P., 51n, 61, 65, 96–97, 104n
Cole, Lamont C., 145
Coleridge, Samuel T., 19, 57
Conservation, 2–4, 65, 70; as anthropocentric, 6, 30; definition of, 5, 101; English laws, 15; splintering of conservation movement, 113, 117, 120n
Cooper, James Fenimore, *The Pioneers*, 21–24
Cowan, Michael H., 34, 51n
Cowdrey, Albert, 16
Crèvecoeur, Hector St. John de, 13, 20
Crockett, Davy, 23
Cronon, William, 10, 15, 25n
Cutright, Paul, 108, 120n

Darwin, Charles, 69, 72–74
DDT, 139–43. See also Carson, Rachel
Deep ecology, 30, 97, 120n, 164; Thoreau as forerunner of, 37
Deforestation, 14, 17, 59, 115; as ecological warfare, 13, 112; in *The Pioneers*, 22–23
Devall, Bill, 97–98, 103n, 120n, 164. See also Deep ecology
Dillard, Annie, 174
Douglas, William O., 142, 169–70
Downing, Andrew Jackson, 49
Dubos, Rene, 124
Dwight, Timothy, 17, 19

Earth First! 162–63, 165n
Easterbrook, Gregg, 171
Eastlake, William, 162
Ecology, 3, 15, 56
Edison, Thomas, 69

Edwards, Jonathan, 18
Elder, John, 86, 96
Ellison, Julie, 31, 51n
Emerson, Ralph Waldo, 69, 71–73, 80n, 85, 94–95, 103n; on abolition, 35; "The Adirondacs," 34; anthropocentrism of, 32; enthusiasm for technolgy, 34; funeral oration for Thoreau, 49–50; influence on Thoreau, 37–38; knowledge of natural history, 33; "Nature," 31–32; philosophy of correspondence, 31; relation to "green" politics, 29–30, 33, 35–36; view of nature, 31–33
Environmentalism, 2–4, 75, 107; definition of, 5–6; growth of advocacy groups, 168; public support for, 168. See also environmental ethics
Environmental ethics, 2–3, evolution of, 30–31, 35
Environmental Protection Agency, 170, 173
Environmental reform, 2–4, 21, 26n, 29–31, 69, 85, 102, 125; dangers of overreliance on litigation, 170–71; difficulty of changing environmental policy, 167–68; first vs. second generation issues, 137; importance of nature writers to, 168–69; opponents to, 172–73; professionalization of, 169; reformers forced to use language of resource economists, 97–98; rhetorical excess used in furthering, 171–72; success in passing landmark legislation, 168
Exxon Valdez, 172

Farquhar, Francis, 94, 103n
Favre, David F., 169
Firestone, Harvey, 69
Fletcher, Ryland, 55
Foerster, Norman, 81n, 85
Ford, Henry, 69
Foreman, Dave, 124, 162
Forest Reserve Act of 1891, 100
Fox, Stephen, 95, 96
Franklin, Benjamin, 17, 19
Freneau, Philip, 20
Frontier, 15, 42, 107–108, 112

Garrison, William Lloyd, 35
Genesis 1:28, 12, 18, 30, 41, 61

Index

Greeley, Horace, 49
Grinnell, George Bird, 112

Haeckel, Ernst, 56
Harding, Walter, 37, 46
Hariot, Thomas, 11
Hays, Samuel P., 3
Haywood, Bill, 162
Helvarg, David, 173
Hetch-Hetchy Valley, 101–102, 160. *See also* Muir, John
Higginson, Francis, 11–12
Hitchcock, Ethan, 101
Hoag, Ronald Wesley, 44
Hoagland, Edward, 175
Hochbaum, Albert, 129
Howells, William Dean, 63; review of Burroughs's *Wake-Robin*, 68
Hudson River School, 26n
Hull, John, 13
Humboldt, Alexander von, 97
Huth, Hans, 19, 26n, 30, 51n, 57

Irving, Washington, 21, 24

Jackson, Andrew, 24, 26n
Jamestown colony, 11
Jefferson, Thomas, 19, 25n, 26n, 158; nature as symbol for America, 20; *Notes on the State of Virginia*, 15, 18–20
Johnson, Edward, 14, 25n
Johnson, Robert, 11
Johnson, Robert Underwood, 85
Josselyn, John 16

Kalm, Peter, 16–19
Kant, Emmanuel, 57
King, Clarence, 94
King Philip's War, 12–13
Krutch, Joseph Wood, 152

Land Ethic, 2, 124. *See also* Leopold, Aldo
Lapham, Increase S., 97
Leatherstocking Tales (Cooper), 21–24
Leopold, Aldo, 2, 4, 23, 60, 63, 110, 138, 143, 153, 159, 164n, 168–69, 171, 174; as bridge between factions of conservation movement, 126, 133; early career of, 126–27; on ecology and habitat protection, 127; as eco-prophet, 123–24; intended audience of, 124–25, 133; interest in hunting and outdoor life, 125, 127, 133; "land ethic" of, 124–25, 128, 130–33; rhetorical technique of, 127, 129–31; *A Sand County Almanac*, 2, 63, 123–24, 129–30; on Theodore Roosevelt, 120
Lincoln, Abraham, 24
Lincklaen, John, 56
Literary nationalism, 20–21
Loeffler, Jack, 162
London, Jack, 119, 120n
Long, William, 70, 119
Lopez, Barry, 2–3, 6, 175
Lowell, James Russell, 37, 39, 45, 51n, 72
Lowenthal, David, 60
Lujan, Manuel, 172

McCann, Garth, 157, 165n
McCay, Mary, 142
McCloskey, H. J., 171
McFarland, J. Horace, 118
McIntosh, James, 36–37, 44, 48
McKibben, Bill, 69, 80
McKinley, William, 106–107
McLane Bird Protection Bill (1913), 78
McPhee, John, 174
Marsh, George Perkins, 2, 40, 85, 110, 143; anthropocentrism of, 56, 60–61, 64; calls for creation of forest preserves, 59; cited in support of creation of Yellowstone and Adirondack Parks, 64; comparison with Pinchot, 65, 113; difficulty of his prose, 63–64; on effects of deforestation, 59–60, 62; impact on environmental reform, 56; influence of transcendentalism on, 57; influence on John Muir, 97–98; *Man and Nature*, 55–58, 97, 114–15; political career, 57–58, 61; precursor of ecology, 60–61; on reforestation, 61–63; *Report on the Artificial Propagation of Fish*, 55, 66; witness to deforestation of Vermont, 56–57
Marsh, James, 57
Marx, Leo, 19, 25
Mather, Cotton, 1, 15, 19
Matthiessen, Peter, 174
Meeker, Joseph W., 1
Meine, Curt, 126–27
Miller, Charles A., 20, 25n

Mitchell, Lee Clark, 10, 24, 26n, 107
Mitchell, Robert Cameron, 137
Monkey wrenching, Thoreau and, 40; traced to Luddites, 159–60. *See also* Abbey, Edward
Morris, Clarence, 169
Morton, Thomas, 11
Mourt's Relation, 11, 25n
Muir, Daniel, 86–89, 102n
Muir, David, 89–90
Muir, John, 18, 45, 49–50, 61, 110, 123, 153, 168, 174; attacked by opponents of wilderness preservation, 146; biocentrism of, 85, 88, 91–92, 97, 102; compared to Edward Abbey, 158–59, 163; compared to John Burroughs, 68, 70, 78–80; contribution to wilderness preservation movement, 85, 96–100, 102; fights to save Hetch-Hetchy Valley, 101–102; glacial theories of, 93–94; influence of George Perkins Marsh, 64; influence of transcendentalism, 94–95; *My First Summer in the Sierras*, 93; pantheism of, 84–85, 90–91, 95–96; as polemicist for environmental reform, 2, 4, 85; rebels against father's religious fundamentalism, 86–89, 102n; rhetorical techniques of, 40, 98–100, 102; and Theodore Roosevelt, 116–20; and Sierra Club, 85, 102; on sportsman's clubs, 113; *A Thousand-Mile Walk to the Gulf*, 90–92; use of biblical imagery, 90; and wilderness preservation, 5–6; with Emerson in Yosemite, 35
Mumford, Lewis, 56
Murphy v. Butler 140, 142–43

Naess, Arne, 164. *See also* Deep ecology
Nash, Roderick, 29, 42, 49, 51n, 60, 64, 85, 164–65n; on American wilderness, 6n, 9, 11, 24; definition of environmentalism, 5; draws parallel between abolitionism and biocentrism, 35
Native Americans, 1; concept of wilderness, 6n; ecological warfare against, 13, 112; relations with early colonists, 12–13; use of fire to clear forests, 25n
Nature faking, 79, 100, 104n, 119. *See also* Burroughs, John; Roosevelt, Theodore
Nature writing, 1–4, 6, 168–69; modern writers, 174–75; readership for, 175; rhetorical excesses in, 171

Oates, Joyce Carol, 36
Oelschlaeger, Max, 29, 31, 51n
Olmsted, Frederick Law, 49

Paine, Thomas, 3
Parkman, Francis, 24, 112–13, 120n
Passenger pigeons, 16, 22, 77, 79
Paul, Sherman, 34, 44
Penn, William, 17
Pequot War, 12
Pinchot, Gifford, 5, 121n, 126, 169; forest management philosophy of, 114–18; on George Perkins Marsh, 64–65; partnership with Theodore Roosevelt, 118 107, 113; rivalry with John Muir, 100–102
Plymouth colony, 11–12
Porte, Joel, 37
Preservationism, 101, 113; definition of 5–6
Progressivism, 2, 66, 107, 111, 122. *See also* Roosevelt, Theodore
Property rights movement, 172–74
Puritans, 18, 40, 44, 51n; attitudes toward wilderness, 9, 12, 14, 25–26n

Quammen, David, 175

Raynal, Abbe Guillaume Thomas, 20
Renehan, Edward, 69, 71, 80
Ribicoff, Abraham, 149
Richardson, Robert D., Jr., 41
Romanticism, 19, 21
Ronald, Ann, 154, 157, 163, 165n
Roosevelt, Theodore, 2, 23, 69, 96, 100–101, 125, 168; anthropocentrism of, 109; as chronicler of western life, 108–109; conservation policies of, 66, 113–18; founds Boone and Crockett Club, 112; friendship with John Burroughs, 78; hunting of, 110–11, 118; impact on conservation movement, 107, 119–20; influence of Gifford Pinchot on, 114, 118; and nature faking controversy, 119, 120n (*see also* Nature faking); as representative of nineteenth-century attitudes, 112, 120; reputation as environmentalist, 116–17; on value of the

Index

"strenuous life," 107–108, 113–14, 117, 119
Roszak, Theodore, 172

A Sand County Almanac. See Leopold, Aldo
Sayre, Robert, 29
Scheese, Donald, 157–58, 161
Sessions, George, 97–98, 102n, 120n, 164. See also Deep ecology
Seton, Ernest Thompson, 70
Sharp, Dallas Lore, 70
Shepard, Odell, 47
Sierra Club, 49, 85, 102, 160
Sierra Club v. Morton, 169
Silent Spring. See Carson, Rachel
Slovic, Scott, 51n, 165n
Spock, Marjorie, 143
Smith, Herbert F., 98
Smith, John, 10
Snyder, Gary, 174
Spotted owl, 167, 173
Stegner, Wallace, 124
Stone, Christopher, 159, 169–70
Stowe, Harriet Beecher, 3, 137
Strachey, William, 11
Swedenborg, Emmanuel, 31

Taft, William Howard, 101, 117
Tallmadge, John, 85, 93, 130–31
Taylor, Paul, 3
Taylor, Zachary, 58
Thoreau, Henry David, 2, 4, 18, 69–70, 79, 85, 119, 123; advocates creation of forest preserves, 43, 46, 48–49, 116; anticipation of "monkey wrenching," 40; antislavery sentiments, 49; "Chesuncook," 39, 45–46, 48, 52n, 63, 90; compared to Edward Abbey, 153, 158–59; compared to John Burroughs, 75, 77; compared to John Muir, 92; ecological awareness of, 38, 41, 46–48, 51n; Emerson's influence on, 35, 37; as forerunner of deep ecology, 37; and "green" politics, 29–30, 33; opinions on hunting, 45–46, 52n; identifies God with Nature, 38; influence on George Perkins Marsh, 57, 63, 66n; influence on John Muir, 94, 103n; "Ktaadn," 43–44; lack of political activity, 49–50; *The Maine Woods,* 43, 52n; opinions on logging, 44–45; pantheism of, 38–39, 41; rejection of anthropocentrism, 39–40, 45, 49; on trans-Atlantic cable, 34; *Walden,* 36, 38, 41–42, 46, 49, 52n; *A Week on the Concord and Merrimack Rivers,* 36, 38–41, 49, 52n; wildness as complement to civilization, 42–43; work on seed dispersal and forest regeneration, 36–37, 47–48
Tichi, Cecelia, 25–26n, 30, 42, 63
Tocqueville, Alexis de, 20–21
Transcendentalism, 29–30, 57, 69, 72, 84. See also Emerson, Ralph Waldo; Thoreau, Henry David
Turner, Frederick, 42, 108, 125

Udall, Stewart, 35, 137

Very, Jones, 57

Watt, James, 172
Westbrook, Perry, 69, 81n
White, Lynn, Jr., 30
Whitman, Walt, 21, 31, 36, 69, 168; influence on John Burroughs, 72–74, 81n
Whitney, Josiah D., 94
Whitten, Jamie, 137
Wild, Peter, 153, 157
Wilderness, as defining aspect of North America, 1; early reactions to, 9–10, 13, 17; preservation of, 86, 97, 111
Williams, Roger, 13, 25n
Wilson, Alexander, 19, 108
Wilson, Woodrow, 102
Winslow, Edward, 11
Winthrop, John, 17
"Wise use," 6, 101, 107, 126, 172–74
Wolfe, Linnie Marsh, 96, 103n
Wood, William, 13; *New England's Prospect,* 11, 47
Wordsworth, William, 96
Worster, Donald, 29, 34, 47, 51n, 65, 107, 111

Yaffee, Steven Lewis, 167

UNIVERSITY PRESS OF NEW ENGLAND
publishes books under its own imprint and is the publisher for
Brandeis University Press, Dartmouth College, Middlebury College Press,
University of New Hampshire, University of Rhode Island, Tufts University,
University of Vermont, Wesleyan University Press, and Salzburg Seminar.

Library of Congress Cataloging-in-Publication Data
Payne, Daniel G., 1958–
Voices in the wilderness : American nature writing and environmental politics /
Daniel G. Payne.
p. cm.
Revision of thesis (doctoral)—State University of New York at Buffalo.
Includes bibliographical references and index.
ISBN 0-87451-751-6 (cloth : alk. paper). — ISBN 0-87451-752-4 (paper : alk. paper)
1. Environmentalism—United States—Authorship. 2. United States—Environmental conditions—History. 3. Environmentalists—United States. 4. Philosophy of nature.
I. Title.
GE197.P39   1996
333.7'2'092273—dc20   95-39422